Principles of Crop Improvement
At a Glance

The Author

Dr. Phundan Singh, hails from a reputed agricultural family of Western Uttar Pradesh [Village-Sakauti, Tehsil- Mawana, District- Meerut]. He graduated in Agriculture from Agra University, Agra and obtained his Master and Doctorate degrees from Kanpur University, Kanpur. He has throughout a brilliant academic record. His area of specialization is Plant Breeding and Genetics. Dr. P. Singh has over 40 year experience in the field of Plant Breeding and Genetics. He has served the Central Institute for Cotton Research, Nagpur for about 32 years in various capacities such as Scientist, Senior Scientist, Principal Scientist, Head of Division and Director. He has 300 publications to his credit. He has participated and presented 80 research papers in various National and International Seminars, Symposia and Conferences. He has authored more than 80 books related to Plant Breeding and Genetics. He has also authored 18 technical bulletins on various aspects of cotton and contributed 14 chapters in various books. Dr P. Singh is a member of several Scientific Societies, referee of different journals, Expert member of Selection Committees and M. Sc. and Ph. D. Examiner in different Agricultural Universities. The Institute was recognized as the referral Laboratory for detection of Bt. gene and received the best Annual Report Award in 2003 when he was the Director. He was conferred with Lifetime achievement Award in 2018 by Cotton Research and Development Association, Hisar. He has visited several countries such as Russia, Belarus, Ukraine, Crimea, Uzbekistan, Canada, USA, England, Austria and Germany.

Principles of Crop Improvement *At a Glance*

[As per 5th Deans' Committee of ICAR]

— *Author* —

Phundan Singh

Former Director
ICAR-Central Institute for Cotton Research,
Nagpur – 440 010
Maharashtra, India

2025

Astral Cracker

A Division of

Astral International Pvt. Ltd.

New Delhi – 110 002

ISBN: 978-93-5919-604-6

Published by : **Astral Cracker**
A Division of
Astral International Pvt. Ltd.
– ISO 9001:2015 Certified Company –
4736/23, Ansari Road, Darya Ganj,
New Delhi-110 002
Ph. 011-43549197, 23278134
E-mail: info@astralint.com
Website: www.astralint.com

— Dedicated to —

Shalini Singh [Daughter-in-law]
Engineer Rajeev Singh [Son]
Dr. Rishika Singh [Granddaughter]
Dr. Muskaan Singh [Granddaughter]

Preface

Plant breeding plays an important role in genetic improvement of crop plants in relation to their economic use for human being. As per 5th Dean Committee Report of ICAR, a new paper entitled "Crop Improvement" has been introduced. There are several books of Plant Breeding. However, there is hardly any book which covers the syllabus of under graduate course related to Crop Improvement. The present book treats the subject in Questions and Answers form to fulfill this long felt need of Indian students.

This book has been designed to covers the undergraduate syllabus of Crop Improvement as per ICAR 5th Dean committee report, 2016. It covers the entire syllabus in one compact volume of 20 chapters. Glossary of technical terms is presented at the end. Some useful information has been appended for general knowledge of students.

The book has been divided into four sections. Section I deals with Introductory Topics, section II with Breeding Methods, III with Seed Production and Crop Breeding and Section IV with Miscellaneous Topics. The book has been written in a simple language for easy grasping of undergraduate students.

The information contained in this book has been gathered from various published sources and Internet websites. Attempts have been made to provide latest information even then some valuable information might have been missed.

The cooperation extended by my wife Mrs. Jaswanti Singh during preparation of this book is praiseworthy. Hope, this Volume would be useful to the students, researchers, and teachers engaged in the field of Crop Improvement. Constructive suggestions of students and teachers are invited for further improvement of this book.

Phundan Singh

Contents

Section III: Seed Production and Crop Breeding

Section IV: Miscellaneous Topics

Syllabus of Crop Improvement I and Crop Improvement II

[As per 5th Deans' Committee of ICAR]

1. Centers of origin, distribution of species, wild relatives in different cereals; pulses; oilseeds; fibres; fodders and cash crops; vegetable and horticultural crops;
2. Plant genetic resources, its utilization and conservation,
3. study of genetics of qualitative and quantitative characters;
4. Important concepts of breeding self-pollinated, cross-pollinated and vegetatively propagated crops;
5. Major breeding objectives and procedures including conventional and modern innovative approaches for development of hybrids and varieties for yield, adaptability, stability, abiotic and biotic stress tolerance and quality (physical, chemical, nutritional);
6. Hybrid seed production technology of Kharif Crops [Maize, Rice, Sorghum, Pearl millet and Pigeon pea, *etc.*],
7. Hybrid seed production of rabi crops
8. Ideotype concept and
9. Climate resilient crop varieties for future.

Section I

Introductory Topics

Chapter 1

Introduction and Objectives

Q1. What is plant breeding?

Ans. Plant breeding is an applied branch of plant genetics which has wide practical application in crop improvement. Plant breeding has been defined in different ways by various authors. Some definitions of plant breeding are given below:

1. J. M. Poehlman (1959) has defined plant breeding as the art and the science of improving the heredity of plants for the benefit of human kind.
2. Frankel (1968) has defined plant breeding as the genetic adjustment of plants to the social, cultural, economic and technological aspects of the environment.
3. In 1978, Riley has defined plant breeding as a technology of developing superior crop plants for various purposes.
4. In 1979, N. W. Simmonds has defined plant breeding as the current phase of crop evolution

Thus plant breeding may be defined as the art, science and technology of improving genetic makeup of crop plants in relation to their economic use for mankind.

Q2. What are other disciplines involved in plant breeding?

Ans. Plant breeding involves several disciplines for development of improved cultivars. The important disciplines which have close relationship with plant breeding and are involved in crop improvement work include: (1) cytogenetics and genetics, (2) morphology and taxonomy, (3) plant physiology, (4) plant pathology, (5) entomology, (6) agronomy and soil science, (7) biochemistry, (8) agricultural engineering, (9) statistics and biometrics, (10) computers and (11) plant biotechnology. Knowledge of all these disciplines is essential for a plant breeder to start a judicious breeding program. Relationship of all these disciplines with plant breeding is presented as follows:

1. **Cytogenetics and Genetics:** Plant breeding is an applied branch of genetics. It involves various genetic principles. Hence, knowledge of cytogenetics and genetics is essential to start a crop improvement program.
2. **Morphology and Taxonomy:** Resistance to biotic and abiotic factors is generally associated with some morphological characters. Sometimes, resistant genes are found in wild relatives of crop plants. Moreover, development of ideal plant type has some morphological bases. Hence some knowledge of plant morphology and taxonomy is essential to a plant breeder for utilization of morphological variation and transfer of resistant genes from wild sources.
3. **Plant Physiology:** Crop plants suffer from abiotic stresses, *viz.* drought, salinity, heat and cold. Development of varieties for such situations requires some knowledge of plant physiology. Moreover, sometimes breeder has to develop physiologically efficient genotypes through the exploitation of genetically controlled physiological variation. For example, there are some physio logical parameters, *viz.* (1) high CO_2 fixation efficiency, (2) high translocation efficiency, (3) high nutrients absorption capacity, (4) low transpiration rate, (5) low photorespiration, (6) high harvest index, (7) high sink capacity, (8) photo-insensitivity, and (9) thermo-insensitivity, which help in promoting crop yield and production. Utilization of these physiological parameters in plant breeding requires a close cooperation of plant physiologist or some training in plant physiology.
4. **Plant Pathology:** Crop plants are infested by a number of fungal, bacterial and viral diseases. Plant breeder has to evolve resistant varieties for various diseases for which some knowledge of plant pathology is essential. Moreover, close association of plant pathologist is essential for achieving such goals.
5. **Entomology:** Crop plants are also attacked by large number of insect species. In some cases, insects pose serious threat to the cultivation of a plant species. In such situation, genetic resistance is the only answer. For development of insect resistant or tolerant cultivars, a plant breeder needs close cooperation of entomologist or some training in entomology.
6. **Agronomy and Soil Science:** Knowledge of the principles of crop production is essential to raise a good crop. This helps in selection and proper evaluation of breeding material. The material should be evaluated under optimum condition. Sometimes, the breeder has to evolve crop varieties resistant to salinity or acid sulphate soils. The resistant lines are identified and isolated by screening the available germplasm on saline or acidic soils. Accomplishment of such task

requires close cooperation of soil scientist or some training in soil science.

7. **Biochemistry:** Various chemical tests are conducted to determine protein, amino acids, oil and fatty acid contents especially in food crops. Moreover, chemical tests are also conducted to determine the presence of toxic compounds in crop like *Khesari dal* (*Lathyrus sativus*), brassica and cotton seed. Such task requires close cooperation of a biochemist or some knowledge of biochemistry.
8. **Agricultural Engineering:** Use of machinery in crop harvesting requires development of varieties with adaptation to mechanical harvesting. In wheat, sorghum, berseem and many vegetable crops including tomato and potato, varieties are available which can be harvested with machine. Development of crop varieties suitable for mechanical harvesting needs close cooperation between agricultural engineer and plant breeder.
9. **Statistics and Biometrics:** Plant breeder has to test the performance of various breeding materials in field experiments. Hence knowledge of experimental designs and statistical methods is essential for a plant breeder. He has to carryout lot of biometrical analysis, *viz.* correlations, path coefficient, discriminant function, diallel, partial diallel, line x tester analysis, stability analysis, D^2 statistics, *etc.* Hence he should be well versed with various biometrical techniques.
10. **Computer:** It is difficult to carryout various biometrical analyses with simple calculators. Now computer programs are available for various biometrical analyses. The calculations which require months together for completion can be finished in few minutes through computers. Moreover, computers are useful in compiling information and plant modeling studies. Thus some knowledge of computer handling is essential to a plant breeder.
11. **Plant Biotechnology:** Plant biotechnology is a combination of plant tissue culture and genetic engineering. Biotechnology is useful tool for development of transgenic crop plants with herbicide resistance, good quality and resistance to biotic and abiotic stresses. It also makes distant crosses possible through somatic hybridization. Thus knowledge of plant biotechnology is essential for a plant breeder.

Q3. What are objectives of plant breeding?

Ans. In plant breeding, there are two types of objectives, *viz.*, general objectives and crop specific objectives. A brief account of some general objectives of plant breeding is presented below:

1. **Higher Yield:** The main objective of a plant breeder is to improve the yield of economic produce which differs from crop to crop. Improvement in yield can be achieved either by evolving high yielding varieties or hybrids.

2. **Improvement in Quality:** The price of produce is determined by its quality. Again quality differs from crop to crop. It refers to cooking quality in rice, baking quality in wheat, malting quality in barley, fibre length, strength and fineness in cotton, nutritive and keeping quality in fruits and vegetables, protein content in pulses, oil content in oil-seeds and sugar content in sugarcane and sugar beet, *etc.*
3. **Resistance to Biotic Stress:** Biotic stress refers to stress caused by biotic factors such as insects, diseases and parasitic weeds. In crop plants considerable yield losses are caused by insects and diseases. Genetic resistance is the cheapest and the best method of minimizing such losses. Resistant varieties are developed through the use of resistant donor parents available in the gene pool.
4. **Resistance to Abiotic Stress:** Crop plants also suffer from abiotic factors such as drought, soil salinity, heat, wind, cold and frost. Breeder has to develop resistant varieties for such environmental conditions.
5. **Wider Adaptability:** Adaptability is an important objective in plant breeding because it helps in stabilizing the crop production over regions and seasons. Adaptability refers to suitability of a variety for general cultivation over a wide range of environmental conditions.
6. **Early Maturity:** Early maturity is a desirable character which has several advantages. It requires less crop management period, less insecticidal sprays, permits double cropping system and reduces overall production cost. Thus earliness is an important objective in plant breeding programs. Determinate growth habit has close association with earliness.
7. **Insensitivity to Temperature and Light:** Development of varieties insensitive to light and temperature helps in crossing the cultivation boundaries of crop plants. In maize, rice and potato now varieties are available which can be grown during summer as well as winter season. Evolution of photo and thermo-insensitive varieties permits their cultivation in new areas outside the boundaries of cultivation of a crop species.
8. **Synchronous Maturity:** It refers to maturity of a crop species at one time. This character is highly desirable in crops like green gram, cowpea, and cotton where several pickings are required for crop harvest.
9. **Desirable Agronomic Characters:** It includes plant height, branching, tillering capacity, growth habit *etc.* Usefulness of these traits also differs from crop to crop. For example, tallness, high tillering and profuse branching are desirable characters in fodder crops, whereas dwarfness is a desirable character in wheat, rice, sorghum and pearl millet. Dwarfness confers lodging resistance in these field crops, in addition to better fertilizer response.

10. **Development of Toxin Free Varieties:** It is essential to develop varieties free from toxic compounds in some crops to make them safe for human consumption. For example, removal of neurotoxin in khesari, which leads to paralysis of lower limbs, erucic acid from Brassica which is harmful for human health, and gossypol from the seed of cotton is necessary to make them fit for human consumption.

11. **Crop Specific Objectives:** In some crops such as green gram, black gram and pea, seeds germinate in the standing crop before harvesting if rains are received. A period of dormancy has to be introduced in these crops to check loss due to germination. In arboreum cotton shedding of kapas after boll bursting is a serious problem. Locule retentive varieties have to be developed in this species of cotton. The shattering of pods is a serious problem in green gram. Hence resistance to shattering is an important objective in green gram.

Q4. What are main activities of plant breeding?

Ans. As stated above, plant breeding started millions of years ago with human civilization. Development of new varieties that are higher-yielding, resistant to pests and diseases, drought-resistant or regionally adapted to different environments and growing conditions is important for ensuring food security. There are several activities of plant breeding. The important activities of plant breeding are listed below:

(i) Creation of variability through hybridization, tissue culture and mutagenesis,

(ii) Selection of elite types from the diverse populations,

(iii) Evaluation of selected genotypes in multi-location trials,

(iv) Identification of superior genotype based on testing,

(v) Seed multiplication, and

(vi) Distribution of the new variety to the farmers for commercial cultivation.

Q5. What are plant breeding organizations in India?

Ans. Now plant breeding is practiced worldwide by individuals such as gardeners and farmers, or by professional plant breeders employed by organizations such as government institutions, universities, crop-specific industry associations or research centers. In each country, there are two types of organization, *viz.* public sector institutions and private sector institutions, where plant breeding work is carried out.

Public Sector Organizations: Such organizations include following.

(i) State Agricultural Universities,

(ii) Central Agricultural Universities,

(iii) ICAR crop research institutes/centers

(iv) State departments of agriculture, and

(v) International crop research institutes/centers

There are about 72 state and central agricultural universities, equal number of ICAR institutes and several International crop research institutes where plant breeding work is carried.

Private Sector Organizations: Such organizations include following.

(i) National private seed companies,

(ii) Multinational private seed companies,

(iii) Farmers and gardeners

Large number of national and International private seed companies are engaged in the plant breeding work. List of some Indian seed companies is appended at the end of the book.

Q6. What are the characteristics improved by plant breeding?

Ans. The significant practical achievements of plant breeding include (i) improvement in yield, (ii) improvement in quality, (iii) insect and disease resistant varieties, (iv) lodging resistant varieties, (v) early maturity, and (v) adaptability.

(i) Improvement in Yield: The genetic yield potential has been improved in several crop plants. The yield of wheat, rice, sorghum and pearl millet has been doubled throughout the world due to improvement in harvest index. The concept of crop ideotype has also helped in building up of ideal plant types for a given set of environmental conditions. However, the concept of plant type has contributed more in cereals than in other crops. Higher yields have also resulted due to exploitation of hybrid vigour in many crops. Heterosis has been successfully exploited in maize, sorghum and pearl millet. The availability of cytoplasmic exploited in maize, sorghum and pearl millet. The availability of cytoplasmic male sterility has facilitated the production of hybrid seed in these crops.

(ii) Improvement in Quality: Quality improvement has been achieved in several crop plants. For example, elimination of the toxic substance erucic acid from the oil of brassica, entirely through genetic means has greatly enhanced the value of this crop (rape and mustard) as a source of edible oil. Presence of neurotoxin in khesari dal (*Lathyrus sativus*) seed has toxic effects on human healht, which results in paralysis of lower limbs called lathyrism. Now varieties of lathyrus have been developed from IARI which have neurotoxin content below the critical level. Varieties with high sugar content in sugarcane and sugar beet, high oil content in oil seed crops and high protein content in pulse crops have been released. In fruit and vegetable crops, varieties with attractive features and good keeping quality have been developed.

(iii) **Insects and Disease Resistant Varieties:** In many crops damage due to insects and diseases has been substantially reduced by developing resistant varieties. Moreover, protection from diseases has also been provided through the development of multiline cultivars in wheat, barley and oats and synthetic cultivars in maize. Similarly, varieties tolerant to drought and salinity have been developed in many field crops.

(iv) **Lodging Resistant Varieties**: Like drought and salinity, lodging also used to cause considerable yield losses in many field crops like wheat, rice, sorghum and pearlmillet. Now dwarf varieties with stiff straw have been developed in these crops which can very well withstand lodging. The dwarf varieties can be successfully grown under high fertility conditions without the risk of lodging. shorter and stiffer straw enables the plant to carry a heavy crop and withstand the battering of strong winds and heavy rains. Norin 10 in wheat and Dee-Geo-woo-Gen in rice are the important sources of dwarfing genes. In Castor bean, plant height has been reduced from 10 meters to 1.5 meters. In cereals like wheat and rice, reduced stem height leads to increased development of ear resulting in improved harvest index and higher yields.

(v) **Early Maturity:** Earliness is a desirable character which has several advantages. Early varieties permit multiple cropping system, escape from late season pests, reduce cost on pesticide sprays and management of crop. Maturity duration has been reduced in many crops. For example, maturity has been reduced from 270 days to 170 days in cotton, from 270 days to 120 days in pigeon pea, from 360 days to 270 days in sugarcane and from 270 days to 180 days in castor bean. Development of early maturing varieties in these crops has significantly contributed to increased production, because early varieties can fit well in the multiple cropping systems resulting in increase of overall production.

(vi) **Improvement in adaptability:** The crop production has significantly increased during the past 30 years. This has resulted mainly due to high yield potential and stable performance of newly developed cultivars in various crops. Development of the concept of stability analysis has helped in evaluation of crop varieties in terms of their adaptability. Stability refers to suitability of a variety for general cultivation over a wide range of environmental conditions. Adaptability is assessed through multi-location or multi-season testing. Varieties with wide adaptability have been developed in wheat, rice, sorghum, maize, pearl-millet and many other crops. The adaptation of crops for mechanical harvesting is also important. In cereals like wheat and rice, varieties of uniform height and maturity have been developed which facilitate mechanical harvesting. In USA, cotton varieties which can be mechanically picked have been released for commercial cultivation.

Q7. What are undesirable consequences of plant breeding?

Ans. Plant breeding has several useful applications in the improvement of crop plants. However, there are some undesirable consequences of plant breeding which are listed below:

(i) **Reduction in biodiversity:** Plant breeding results in development of improved varieties which are more uniform than land races. The uniform varieties are more prone to the new races of pathogen than land races which have high genetic diversity.

(ii) **Narrow genetic base:** Improved varieties have narrow genetic base and poor adaptability than land races which have broad genetic base and wide adaptability.

(iii) **Danger of uniformity:** Improved varieties have some common parents in the pedigree which may cause danger of uniformity.

(iv) **Useless combinations:** Sometimes, plant breeding leads to undesirable combination of characters. Examples of undesirable combination of characters are *Raphanobrassica* and Pomato.

Q8. What are future prospects of plant breeding?

Ans. Plant breeding has played key role in increasing the yield levels of cereals, millets and many other crops. In future, plant breeding efforts need to be directed towards the following thrust areas.

1. There is ample scope for genetic improvement of pulses and oilseed crops.
2. In India, considerable area is under rain-fed cultivation. Hence, there is need to develop crop cultivars resistant to drought conditions.
3. In view of global warming, there is need to develop crop cultivars suitable for climate change.
4. The labor has become very expensive. Hence, there is need in many crops to develop varieties suitable for machine harvesting.
5. The problem of metal toxicity is increasing in many areas particularly adjacent to industries. Hence, there is need to develop crop cultivars tolerant to metal toxicity.

Now new plant breeding approaches, *viz.*, molecular breeding, somatic hybridization and genetic engineering are available. In future, molecular breeding and biotechnology will supplement the conventional plant breeding methods for genetic improvement of crop plants.

Chapter 2

Centres of Origin and Distribution of Crop Species

Q1. What is genetic diversity?

Ans. Genetic diversity is the total amount of genetic variation present in a population or species. In other words, crop genetic diversity refers to the variety of genes and genotypes found in a particular crop species. Genetic diversity is essential to develop improved cultivars with broad genetic base and wide adaptability. Moreover, existence of genetic diversity is also essential to meet current and future breeding requirements. If a crop species has large number of genetic variants it is said to be genetically diverse. Genetic diversity provides broad genetic base to a population. Genetic diversity is depleted due to genetic erosion and extinction. Therefore, collection of germplasm is essential to conserve the genetic diversity and to minimise its loss due to genetic erosion and extinction.

Q2. Define center of diversity.

Ans. Centre of diversity refers to the geographic region in which greatest variability of a crop occurs. A primary center of diversity is the region of presumed origin, and secondary centers of diversity are regions of high diversity which have developed as a result of subsequent spread of a crop. The main differences between centers of origin and centers of diversity are given below"

(i) Centers of origin are geographical areas where crop plants have originated.

(ii) A center of diversity refers to a location where vast genetic variability for a crop and its wild species is found.

Thus, the center of origin and center of diversity for a crop may the same or may be different (Fehr, 1987). Vavilov could not adequately cover

Africa. Moreover, Australia was not covered. These two continents have tremendous wealth of crop genetic diversity of several crop plants.

Q3. What are vavilovian centers of diversity?

Ans. N.I. Vavilov (1926, 1951), a Russian geneticist and plant breeder, was the pioneer man who realized the significance of genetic diversity for crop improvement. Vavilov and his colleagues visited several countries and collected cultivated plants and their wild relatives for use in the Russian breeding program to develop varieties for various agro-climatic conditions of USSR. Based on his studies of global exploration and collection, Vavilov proposed eight main centers of diversity and three subsidiary centers of diversity given as follows:

Q4. What are main centers of diversity?

Ans. Main centres of crop diversity as proposed by Vavilov are: (1) China, (2) India (Hindustan), (3) Central Asia, (4) Asia Minor or Persia, (5) Mediterranea, (6) Abyssinia, (7) Central America or Mexico, and (8) South America.

1. **China:** Naked oat (SC), Soybean, Adzuki bean, Common bean (SC), Small Bamboo, Leaf Mustard (SC), Apricot, Peach, Orange, Sesame (SC), China tea, *etc.*
2. **Hindustan** : Rice, Chick Pea, Moth Bean, Rice bean, Horse gram, Brinjal, Cucumber, Tree Cotton, Jute, Pepper, African Millet, Indigo, *etc.*
3. **Central Asia**: Bread wheat, Club wheat, Shot wheat, Rye (SC), Pea, Lentil, Chickpea, Sesame, Flax, Safflower, Carrot, Radish, Apple, Pear and Walnut.
4. **Asia Minor Persia**: Einkorn wheat, Durum wheat, Poulard wheat, Bread wheat, Two Rowed barley, Rye, Red oat, Chickpea (SC) lentil, Pea (SC), Flax, Almond, Pomegranate, Pistachio, Apricot and Grape.
5. **Mediterranea**: Durum wheat, Husked oats, Cabbage, Olive, Broad bean and Lettuce.
6. **Abyssinia**: Durum wheat, Poulard wheat. Emmer wheat, Barley, Chickpea, Lentil, Pea, Flax. Sesame, Castor bean, African Millet, and Coffee.
7. **Central America**: Maize, Common bean, Upland cotton, Pumpkin, Gourd, Squash, Sisal or Mexico hemp and Pepper.
8. **South America**: Potato, Sweet potato, Lima bean, Tomato, Papaya, Tobacco and Sea Island cotton.

Q5. What are subsidiary centers of diversity?

Ans. There are three subsidiary centers of diversity. These are: (1) Indo- Malaya, (2) Chile, and (3) Brazil and Paragua. All these centers are known as centers of origin or centers of diversity or Vavilovian centers of diversity.

1. **Indo-Malaysia**: Banana, Coconut, Yam, and Pomelo
2. **Chile**: Potato.
3. **Brazil and Paraguay**: Peanut, Rubber Tree, Cocoa (SC), Fineapple, *etc.*

Q6. What are types of center of diversity?

Ans. The centers of crop diversity are of three types *viz.* (1) primary centers of diversity, (2) secondary centers of diversity, and (3) micro-centers. These are briefly discussed below:

(1) Primary Centers of Diversity

Primary centers are regions of vast genetic diversity of crop plants. These are original homes of the crop plants which are generally uncultivated areas like, mountains, hills, river valleys, forests, *etc.* Main features of these centers are given below:

(i) They have wide genetic diversity.
(ii) Have large number of dominant genes.
(iii) Mostly have wild characters.
(iv) Exhibit less crossing over.
(v) Natural selection operates.

(2) Secondary Centers of Diversity

Vavilov suggested that valuable forms of crop plants are found far away from their primary area of origin, which he called secondary centers of origin or diversity. These are generally the cultivated areas and have following main features:

(i) Have lesser genetic diversity than primary centers.
(ii) Have large number of recessive genes.
(iii) Mostly have desirable characters.
(iv) Exhibit more crossing over.
(v) Both natural and artificial selections operate.

(3) Micro-centers

In some cases, small areas within the centers of diversity exhibit tremendous genetic diversity of some crop plants. These areas are referred to as micro-centers. Micro-centers are important sources for collecting valuable plant forms and also for the study of evolution of cultivated species. The main features of micro centers are given below:

(i) They represent small areas within the centers of diversity.
(ii) Exhibit tremendous genetic diversity.
(iii) The rate of natural evolution is faster than larger areas.
(iv) They are important sites for the study of crop evolution.

Q7. What is law of parallel variation?

Ans. The concept of parallel variation also known as law of homologous series of variation was developed by Vavilov (1951) based on his study of crop diversity and centers of origin. Law of homologous series states that a particular variation observed in a crop species is also expected to be available in another related species. For instance, if we get dwarf collections in one species of a crop, the same may be observed in another related species also. Vavilov used principle of homologous series of variation as a clue for discovering similar characters in related species.

Q8. What are gene sanctuaries?

Ans. The genetic diversity is sometimes conserved under natural habitat. In other words, areas of great genetic diversity are protected from human interference. These protected areas in natural habitat are referred to as gene sanctuaries. Gene sanctuary is generally established in the center of diversity or micro-center. Gene sanctuary is also known as natural park or biosphere reserve. India has setup its first gene sanctuary in the Garo Hills of Assam for wild relatives of citrus. Efforts are also being made to setup gene sanctuaries for banana, sugarcane, rice and Mango. In Ethiopia gene sanctuary for conservation of wild relatives of coffee was setup in 1984. Gene sanctuaries have two main advantages. Firstly, it protects the loss of genetic diversity caused by human intervention. Secondly, it allows natural selection and evolution to operate. There are two main drawbacks of gene sanctuary. Firstly, entire variability of a crop species can not be conserved. Secondly, its maintenance and establishment is a difficult task. It is a very good method of *in situ* conservation.

Q9. What are gene banks?

Ans. Gene bank refers to a place or organization where germplasm can be conserved in living state. Gene banks are also known as germplasm banks. The germplasm is stored in the form of seeds, pollen or *in vitro* cultures, or in the case of a field gene banks, as plants growing in the field. Gene banks are mainly of two types, *viz.* (1) seed gene banks, and (2) field gene banks. These are briefly discussed below :

1. **Seed Gene Bank:** A place where germplasm is conserved in the form of seeds is called seed gene bank. Seeds are very convenient for storage because they occupy smaller space than whole plants. However, seeds of all crops can not be stored at low temperature in the seed banks. The germplasm of only orthodox species (whose seed can be dried to low moisture content without losing variability) can be conserved in seed banks. In the seed banks, there are three types of conservation, *viz.* (1) short term, (2) medium term, and (3) long-term. Base collections are conserved for long term (50 years or more) at –18 or –20°C. Active collections are stored for medium term (10-15 years) at zero degree celsius and working collection are stored for short term (3-5 years) at 5-10°C. The main advantages of gene banks are as follows:

(1) Large number of germplasm samples can be conserved in a very small space.

(2) In seed banks, handling of germplasm is easy.

(3) Germplasm is conserved under pathogen and insect free environment.

There are some disadvantages of germplasm conservation in the seed banks. These are listed below:

(1) Seeds of recalcitrant species can not be stored in seed banks.

(2) Failure of power supply may lead to loss of viability and thereby loss of germplasm.

(3) It requires periodical evaluation of seed viability. After some time multiplication is essential to get new or fresh seeds for storage.

2. **Field Gene Banks:** Field gene banks also called plant gene banks are areas of land in which germplasm collections of growing plants are assembled. This is also called *ex-situ* conservation of germplasm. Those plant species that have recalcitrant seeds or do not produce seeds readily are conserved in field gene banks. In field gene banks, germplasm is maintained in the form of plants as a permanent living collection. Field gene banks are often established to maintain working collections of living plants for experimental purposes. They are used as source of germplasm for species such as coconut, rubber, mango, cassava, yam and cocoa. Field gene banks have been established in many countries for different crops.

Field gene banks in some countries

1. **Malaysia:** Oil palm has been conserved on 500 hectares.
2. **Indonesia:** Earmarked 1000 hectare area for coconut and other perennial crops.
3. **Philippines:** South East Asian germplasm of banana has been conserved.
4. **India:** Global collection of coconut has been conserved in Andman and Nicobar.

Field gene banks have some advantages and disadvantages which are discussed below: There are two main advantages.

1. It provides opportunities for continuous evaluation for various economic characters.
2. It can be directly utilized in the breeding program

There are three main demerits of field gene banks as given below:

1. Field gene banks can not cover the entire genetic diversity of a species. It can cover only a fraction of the full range of diversity of a species.
2. The germplasm in field gene banks is exposed to pathogens and insects and sometimes is damaged by natural disasters such as bushfires, cyclones, floods, *etc.*
3. Maintenance of germplasm in the field gene banks is costly affair.

Q10. What are meristem gene banks?

Ans. Germplasm of asexually propagated species can be conserved in the form of meristems. This method is widely used for conservation and propagation of horticultural species. *In vitro* method can be used in two ways *viz.* firstly, for storage of tissues under slow growth conditions and secondly, for long term conservation of germplasm by cryopreservation. In cryopreservation, the tissues are stored at a very low temperature *i.e.* at –196oC in liquid nitrogen. At this temperature, all biological processes virtually come to a stop.

Gene Banks for Various Crops in India

1. Wheat: Indian Institute of Wheat and Barley Research, Karnal.
2. Rice: Central Rice Research Institute (CRRI), Cuttack.
3. Potato: Central Potato Research Institute (CPRI), Shimla.
4. Cotton: Central Institute for Cotton Research (CICR), Nagpur
5. Pulses Kanpur Indian Institute for Pulses Research (IIPR)
6. Oilseed crops: Directorate of Oilseed Research (DOR), Hyderabad.
7. Sorghum: Directorate of Sorghum Research, Hyderabad.
8. Soybean: Directorate of Soybean Research, Indore.
9. Groundnut: Directorate of Groundnut Research, Junagarh.
10. Maize: Indian Institute of Maize Research, Pusa, New Delhi.
11. Citrus: Central Citrus Research Institute, Nagpur.
12. Sugarcane: Sugarcane Breeding Institute (SBI), Coimbatore.
13. Forage crops: Indian Grossland and Fodder Research Institute (IGFRI), Jhansi.
14. Tobacco: Central Tobacco Research Institute (CTRI), Rajamundry.
15. Tuber crops: Central Tuber Crop Research Institute (CPCRI), Trivendrum.
16. Horticultural crops: Indian Institute for Horticultural Research (IIHR), Bangalore.
17. Grapes: National Research Centre for Grapes, Pune

Table 2-1. International/Global Gene Banks for different Crops

Sl.No.	*Institute*	*Gene Bank For*
1	International Rice Research Institute (IRRI), Philippines	Rice
2	International Wheat and Maize Improvement Centre (CIMMYT), Mexico	Maize, Wheat, Triticale, Barley
3	International Centre for Tropical Agriculture (CIAT), Colombia	Cassava, Beans, Rice and Maize
4	International Institute for Tropical Agriculture (IITA), Nigeria	Cowpea, Soybean, Limabean, Cassava, Sweet Potato

Sl.No.	*Institute*	*Gene Bank For*
5	International Potato Centre, Peru	Potato
6	International Crop Research Institute for Semiarid Tropics (ICRISAT), Hyderabad, India	Sorghum, Pearl millet, Pigeon pea, Groundnut
7	International Centre for Agriculture Research in Dryland Areas (ICARDA), Syria.	Durum Wheat, Barley and Beans
8	Asian Vegetable Research and Development Centre (AVRDC), Taiwan	Vegetables, Soybean, Mung bean, Sweet potato

Based on status of Research Institutes, gene banks are again of two types, *viz.* (1) National gene banks, and (2) International or global gene banks. National gene banks are maintained by each country and global gene banks are located in International Crop Research Institutes/Centers. In India, gene banks are maintained by concerned Crop Research Institutes of ICAR. National Bureau of Plant Genetic Resources, New Delhi is also maintaining germplasm of various field crops. Germplasm can also be conserved in the form of pollen and DNA. However, these methods are not in common use. Conservation of germplasm in the form of DNA is a difficult task. It requires sophisticated laboratory and is very expensive.

Chapter 3

Plant Genetic Resources: Conservation and Utilization

Q1. Define gene pool and describe its main features.

Ans. The sum total of genes in a crop species is referred to as genetic resources or gene pool or genetic stock or germplasm. In other words, gene pool refers to a whole library of different alleles of a species. Germplasm or gene pool is the basic material with which a plant breeder has to initiate his breeding program. Some important features of plant genetic resources or gene pool are given below:

1. Gene pool represents the entire genetic variability or diversity available in a crop species.
2. Germplasm consists of land races, modern cultivars, obsolete cultivars, breeding stocks, wild forms and wild species of cultivated crops.
3. Germplasm includes both cultivated and wild species or relatives of crop plants.
4. Germplasm is collected from the centres of diversity, gene banks, gene sanctuaries, farmers' fields, markets and seed companies.
5. Germplasm is the basic material for launching a crop improvement program.
6. Germplasm may be indigenous (collected within country) or exotic (collected from foreign countries).

Q2. What are kinds of germplasm?

Ans. The germplasm consists of various plant materials of a crop such as (1) land races, (2) obsolete cultivars, (3) modern cultivars, (4) advanced (homozygous) breeding materials, (5) wild forms (races) of cultivated species, (6) wild relatives, and (7) mutants. These are briefly discussed below:

1. **Land Races:** Land races are nothing but primitive cultivars which were selected and cultivated by the farmers for many generations. Main features of land races are given below:
   (i) Land races were not deliberately bred like modern cultivars. They evolved under subsistence agriculture.
   (ii) Land races have high level of genetic diversity which provides them high degree of resistance to biotic and abiotic stresses. Biotic stress refers to hazards of diseases and insects, whereas abiotic stress means, drought, salinity, cold, frost, *etc.*
   (iii) Land races have broad genetic base which again provides them wider adaptability and protection from epidemic of diseases and insects.
   (iv) Land races even respond to selection for high yield, but to certain extent. Since land races possess valuable alleles, their conservation is essential.
   (v) Land races are less uniform and low yielders. Land races were first collected and studied by N.I. Vavilov in rice. Now land races have been collected in maize, sorghum, pearlmillet and many other crops especially in South Asia.
2. **Obsolete Cultivars:** Improved varieties of recent past are known as obsolete cultivars. These are the varieties which were popular earlier and now have been replaced by new varieties. These varieties have several desirable characters and constitute an important part of gene pool. For example, wheat varieties K68, K65 and Pb 591 were most popular traditional tall varieties before introduction of high yielding dwarf Mexican wheat varieties. These varieties are well known for their attractive grain color and chapati making quality. Now these varieties are no more cultivated. They are good genetic resources and have been widely used in wheat breeding programs especially in India for improvement of grain quality. Now such old varieties are found in the gene pool only.
3. **Modern Cultivars:** The currently cultivated high yielding varieties are referred to as modern cultivars. Modern cultivars are also known as improved cultivars or advanced cultivars. These varieties have high yield potential and uniformity as compared to obsolete varieties and land races. Modern cultivars constitute a major part of working collections and are extensively used as parents in the breeding programs for further genetic improvement in various characters. Hence these cultivars are in great demand. These varieties are the outcome of scientific plant breeding and have been developed for modern intensive agriculture. However, modern cultivars have narrow genetic base and low adaptability as compared to land races.

4. **Advanced Breeding Lines:** Pre-released plants which have been developed by plant breeders for use in modern scientific plant breeding are known as advanced lines, cultures and stocks. They include advanced cultures which are not yet ready for release to farmers. Sometimes advanced breeding lines and stock are not very much productive, but constitute valuable part of gene pool for various economic characters.

5. **Wild forms of Cultivated Species:** Wild forms of cultivated species are available in many crop plants. Such plants have generally high degree of resistance to biotic and abiotic stresses and are utilized in breeding programmes for genetic improvement of resistance to biotic and abiotic stresses. They can easily cross with cultivated species. However, wild forms of many crop species are extinct. Moreover, entire range of diversity of available wild forms is rarely tapped. They constitute small part of gene pool.

6. **Wild Relatives:** Those naturally occurring plant species which have common ancestry with crops and can cross with crop species are referred to as wild relatives or wild species. Wild relatives are important sources of resistance to biotic (diseases and insects) and abiotic (drought, cold, frost, salinity, *etc.*) stresses. However, wild relatives are used as the last resort in crop improvement programs, because their use in crossing leads to : (1) hybrid sterility, (2) hybrid inviability and (3) transfer of several undesirable genes to the cultivated species along with desirable alleles. This group constitutes a minor part of gene pool. Interspecific derivatives are added to the gene pool.

7. **Mutants:** Mutation breeding is used when the desired character is not found in the genetic stocks of cultivated species and their wild relatives. Mutations do occur in nature as well as can be induced through the use of physical and chemical mutagens. The extra variability which is created through induced mutations constitutes important component of gene pool. Mutants for various characters sometimes may not be released as a variety, but they are added in the gene pool. For example, mutant gene pool Dee-Geo-Woo-Gen in rice and Norin 10 in wheat proved to be valuable genetic resources in the development of high yielding and semi dwarf varieties in the respective crop species. In seed propagated crops, 410 varieties have been released through the use of mutants in the crosses (IAEA, 1991).

Q3. Explain briefly the gene pool system of classification.

Ans. Gene pool of a crop includes all cultivars (obsolete and current), wild species and wild relatives containing all the genes available for breeding use. Based on degree of relationship, the gene pool of a crop can be divided into three groups, *viz.* (1) primary gene pool, (2) secondary gene pool, and (3) tertiary gene pool (Harlan and Dewet, 1971). These are briefly discussed below:

1. **Primary gene pool (GP1):** The gene pool in which inter-mating (crossing) is easy and leads to production of fertile hybrids is known as primary gene pool. It includes plants of the same species or of closely related species which produce completely fertile offspring on inter-mating. In such gene pool, genes can be exchanged between lines simply by making normal crosses. This is also known as gene-pool one (GP1). This is the material of prime breeding importance.
2. **Secondary gene pool (GP2):** The genetic material that leads to partial fertility on crossing with GP1 is referred to as secondary gene pool. It includes plants that belong to related species. Such material can be crossed with primary gene pool, but usually the hybrids are sterile and some of the progeny to some extent are fertile. Transfer of gene from such material to primary gene pool is possible but difficult. This type of gene pool is also known as gene-pool two (GP2).
3. **Tertiary gene pool (GP3):** The genetic material which leads to production of sterile hybrids on crossing with primary gene pool is termed as tertiary gene pool or gene pool three (GP3). It includes material which can be crossed with GP1, but the hybrids are sterile. Transfer of genes from such material to primary gene pool is possible with the help of special techniques.

Q4. What are types of seed collections?

Ans. Based on the use and duration of conservation, seed collections are of three types: (1) base collections, (2) active collections, and (3) working collections. These are briefly discussed as follows:

1. **Base Collections:** Base collections include maximum number of accessions (samples) available in a crop. These are meant for long term conservation (up to 50 years or more) and are stored at –18 or –20°C in hermetically sealed containers. The seeds are dried to 5 + 1 per cent moisture and have more than 85 per cent initial seed viability. These collections are disturbed only for the purpose of regeneration. These are used only when germplasm from other sources is not available for use in breeding. It is also known as principal collection and refers to the whole collection.
2. **Active Collections:** This category of germplasm is actively utilized in breeding programs and are conserved for medium term (8-10 years or more). These collections are stored at zero degree Celsius with moisture content around 8 per cent. Germination test is carried out after every 5-10 years to assess the reduction in seed viability.
3. **Working Collections:** There collections are frequently utilized by breeders in their crop improvement programmes. These are stored for short term (3 to 5 years). The seed is stored at 5-10°C with moisture content of 8-10 per cent. There is another category of seed collections called core collection. It refers to a sub-set of base collection which represents the large collection or base collection. In other words, core collection is a limited set of accessions

derived from an existing germplasm collections, chosen to represent the genetic spectrum in the whole collection. The concept of core collection was proposed by Frankel (1984).

Q5. List germplasm activities and describe briefly exploration and collection.

Ans. There are six important activities related to plant genetic resources: *viz.* (1) exploration and collection, (2) conservation (3) evaluation, (4) documentation, (5) distribution, and (6) utilization.

Exploration refers to collection trips and collection refer to tapping of genetic diversity from various sources and assembling the same at one place. The exploration and collection is a highly scientific process. This process takes into account six important items, *viz.* (1) sources of collection, (2) priority of collection, (3) agencies of collection, (4) methods of collection, (5) methods of sampling and (6) sample size. These aspects are briefly discussed below:

1. **Sources of collection:** There are five important sources of germplasm collection : *viz.* (1) centres of diversity, (2) gene banks, (3) gene sanctuaries, (4) seed companies, and (5) farmers' fields. Moreover, collections can be made by local exploration trips to the regions of crop diversity.
2. **Priority of collection:** The next important step in the germplasm collection is to fix priority of collection. Some areas of diversity have been threatened more than others by the danger of extinction. Similarly, some crop species have more risk of extinction than others. Hence, endangered areas and endangered species should be given priority for germplasm collection.
3. **Agencies of collection:** The task of germplasm collection is undertaken by crop research institutes and state agricultural universities in collaboration with National Bureau of Plant Genetic Resources, New Delhi for indigenous collections. For global collection the task is undertaken in collaboration with International Plant Genetic Resources, Institute (IPGRI), Rome, Italy.
4. **Method of collection:** Germplasm collections are made in four principal ways: *viz.* (1) through expeditions to the areas or regions of genetic diversity, (2) by personal visit to gene bank centre, (3) through correspondence, and (4) through exchange of material.
5. **Method of sampling:** There are two sampling methods for collection of germplasm from the regions of diversity, *viz.* (1) random sampling, and (2) biased sampling. Random sampling is effective in capturing of alleles for biotic and abiotic stresses, whereas non- random or biased sampling is useful in collection of morphologically distinct genotypes. Hence, it is advised that both random as well as biased sampling procedures should be adopted to tap the maximum genetic diversity of a crop species.
6. **Sample size:** The sample size should be such that 95 per cent of the total genetic diversity can be captured from the area of collection. To achieve this goal, 50 to 100 individuals should be collected per site with 50 seeds

per plant. A wide range of habitat should be sampled for capturing maximum genetic diversity.

Q6. What are merits and demerits of exploration and collection?

There are several merits and demerits of exploration and collection of germplasm, some of which are as discussed below:

Merits

1. Collection helps in tapping crop genetic diversity and assembling the same at one place. It reduces the loss of genetic diversity due to genetic erosion.
2. Sometimes, we get material of special interest during exploration trips.
3. Sometimes, we come across a new plant species during the process of collection.
4. Collection also helps in saving certain genotypes from extinction. Once the material is collected, it can be maintained further in the germplasm.

Demerits

1. Collection of germplasm especially from other countries, sometimes leads to entry of new diseases, new insects and new weeds.
2. Collection is a tedious job. The collection has to be made generally from uncultivated areas like hills, mountains, river valleys and forests, where the collector faces problems of boarding, lodging and transportation.
3. In the remote areas, the collector, sometimes has encounter with wild animals like elephants, rhinos, tigers, lions and snakes which involves risk of life.
4. Transportation of huge collections also poses difficulties in the exploration and collection.

Q7. Give a brief account of germplasm conservation.

Ans. Conservation refers to protection of genetic diversity of crop plants from genetic erosion. There are two important methods of germplasm conservation or preservation, *viz.* (1) *in situ* conservation, and (2) *ex situ* conservation. These are described below:

1. ***In-situ* Conservation:** Conservation of germplasm under natural habitat is referred to as *in situ* conservation. It requires establishment of natural or biosphere reserves, national parks or protection of endangered areas or species. In this method of conservation, the wild species and the complete natural or semi-natural ecosystems are preserved together. This method of preservation has following main disadvantages.

   (i) Each protected area will cover only very small portion of total diversity of a crop species, hence several areas will have to be conserved for a single species.

(ii) The management of such areas also poses several problems.

(iii) This is a costly method of germplasm conservation.

2. ***Ex-situ* Conservation:** It refers to preservation of germplasm in gene banks. This is the most practical method of germplasm conservation. This method has following three advantages:

(i) It is possible to preserve entire genetic diversity of a crop species at one place.

(ii) Handling of germplasm is also easy.

(iii) This is a cheap method of germplasm conservation.

The germplasm is conserved either (1) in the form of seed, or (2) in the form of meristem cultures. preservation in the form of seed is the most common and easy method. Seed conservation is relatively safe, requires minimum space (except coconut, *etc.*) and easy to maintain. Glass, tin or plastic containers are used for preservation and storage of seeds. The seeds can be conserved under long term (50 to 100 years), medium term (10 to 15 years) and short term (3-5 years) storage conditions.

Q8. What are types of seeds in relation to conservation?

Ans. Roberts (1973) has classified seeds into two groups for storage purpose; *viz.* (1) orthodox and (2) recalcitrant.

1. **Orthodox:** Seeds which can be dried to low moisture content and stored at low temperature without losing their viability are known as orthodox seeds. This group includes seeds of corn, wheat, rice, carrot, beets, papaya, pepper, chickpea, lentil, soybean, cotton, sunflower, various beans, egg plant and all the *brassicas.* These seeds can be dried and stored at low temperatures for long periods of time.

2. **Recalcitrant:** Seeds which show very drastic loss in viability with a decrease in moisture content below 12 to 13 per cent are known as recalcitrant seeds. This group includes cocoa, coconut, mango, tea, coffee, rubber, jackfruit and oil palm seeds. Such seeds cannot be conserved in seed banks and, therefore, require in situ conservation. Crop species with recalcitrant seeds are conserved in field gene banks which are simply areas of land in which collection of growing plants are assembled. For conservation of meristem cultures, meristem or shoot tip banks are established. Conservation of genetic stocks by meristem cultures has several advantages as given below:

(i) Exact genotypes can be conserved indefinitely free from virus or other pathogens and without loss of genetic integrity.

(ii) It is advantageous for vegetatively propagated crops like potato, sweet potato, cassava, *etc.*, because seed production in these crops is poor.

(iii) Vegetatively propagated material can be saved from natural disasters or pathogen attack.

(iv) Long regeneration cycle can be envisaged from meristem cultures.

(v) Perennial plants which take 10 to 20 years to produce seeds can be preserved any time by meristem cultures.

(vi) Regeneration of meristems is extremely easy.

(vii) Plant species having recalcitrant seeds can be easily conserved by meristem cultures.

Q8. Give a brief account of germplasm evaluation.

Ans. Evaluation refers to screening of germplasm in respect of morphological, genetic, economic, biochemical, physiological, pathological and entomological attributes. Evaluation of germplasm is essential from following angles.

(i) To identify gene sources for resistance to biotic and abiotic stresses, earliness, dwarf-ness, productivity and quality characters.

(ii) To classify the germplasm into various groups.

(iii) To get a clear picture about the significance of individual germplasm line.

Evaluation requires a team of specialists from the disciplines of plant breeding, physiology, biochemistry, pathology and entomology. First of all a list of descriptors (characters) for which evaluation has to be done is prepared. This task is completed by a team of experts from IPGRI, Rome, Italy. The descriptors are ready for various crops. The material is evaluated at several locations to get meaningful results. Moreover, evaluation is done in a phased manner. The variation for polygenic characters is assessed by three different methods as given below:

(i) By simple measures of dispersion (range, standard deviation, standard error and coefficient of variation).

(ii) By metroglyph analysis of Anderson (1957), and

(iii) By D_2 statistics of P.C. Mahalanobis (1936).

The evaluation of germplasm is down in three different places, *viz.* (1) in the field, (2) in greenhouse, and (3) in the laboratory. Observations on morphological characters, productivity attributes, resistance to biotic and abiotic stresses, and some physiological parameters like photosynthetic efficiency and transpiration rate can be recorded under field conditions using portable instruments. The resistance to biotic and abiotic stresses can be screened under green-house conditions. Evaluation for biochemical characters like protein, oil and amino acid contents, and technological characters is completed under laboratory conditions. Both visual observations and metric measurements are used for evaluation.

Q10. Explain briefly documentation of germplasm.

Ans. Documentation refers to compilation, analysis, classification, storage and dissemination of information. In plant genetic resources, documentation means dissemination of information about various activities such as collection, evaluation, conservation, storage and retrieval of data. Now the term documentation is more appropriately known as information system. Documentation is one of the important activities of genetic resources. Information system is useful in many ways as given below:

(i) It provides information about various activities of plant genetic resources.

(ii) It provides latest information about characterization, conservation, distribution and utilization of genetic resources.

(iii) It helps explorers, evaluators and curators in the conservation of genetic resources.

(iv) It helps in making genetic resources accessible to plant breeders and other users.

Large numbers of accessions are available in maize, rice, wheat, sorghum, potato and other major crops. About 7.3 million germplasm accessions are available in 200 crops species. Handling of such huge germplasm information is only possible through electronic computers. For uniformity of information IPGRI has designed descriptors (characters) and descriptor state for majority of crops. The entire data is put in the computer memory and the desired information can be obtained any time from the computer.

Q11. Write short note on germplasm distribution.

Ans. The distribution of germplasm is one of the important activities of genetic resources centers. The specific germplasm lines are supplied to the users on demand for utilization in the crop improvement programs.

(i) Distribution of germplasm is the responsibility of the gene bank centers where the germplasm is maintained and conserved.

(ii) The germplasm is usually supplied to the workers who are engaged in the research work of a particular crop species.

(iii) Germplasm samples are generally supplied free of cost to avoid cumbersome work of book keeping.

(iv) The quantity of seed samples to be sent is usually small, depends on the availability of seed material and demands received for the same and several other factors.

(v) Proper records are maintained about the distribution of material. After evaluation users should send a report about important characters of the accessions to the distributor who will record the information in the germplasm register for documentation purpose.

(vi) The germplasm is usually distributed after evaluation by collecting centre for one or two crop seasons. It helps in acclimatization and purification of the material.

(vii) Without distribution to the actual users, there is no point in collecting the germplasm.

Q12. How germplasm is utilized in crop improvement?

Ans. Utilization refers to use of germplasm in crop improvement programs. The germplasm can be utilized in various ways. The uses of cultivated and wild species of germplasm are briefly discussed below :

1. ***Cultivated Germplasm*:** The cultivated germplasm can be used in three main ways : (1) as a variety, (2) as a parent in the hybridization, and (3) as a variant in the gene pool. Some germplasm lines can be released directly as varieties after testing. If the performance of an exotic line is better than a local variety, it can be released for commercial cultivation. In some cases, new variety is developed through selection from the collection. Some germplasm lines are not useful as such, but have some special characters, such as disease resistance, good quality of economic produce, or wider adaptability. These characters can be transferred to commercial cultivars by incorporating such germplasm lines in the hybridization programme. Transfer of desirable characters from cultivated germplasm to the commercial cultivars is very easy because of cross compatibility.
2. ***Wild Germplasm*:** The wild germplasm is used to transfer resistance to biotic and abiotic stresses, wider adaptability and sometimes quality such as fibre strength in cotton. However, utilization of wild germplasm poses three main problems : *viz.* (1) hybrid inviability - inability of a hybrid to survive, (2) hybrid sterility - inability of a hybrid to produce offspring, and (3) linkage of undesirable characters with desirable ones. Thus utilization of wild germplasm for crop improvement is a difficult task.

Q13. What are the organizations associated with germplasm?

Ans. There are two types of organizations, *viz.* International and National which are associated with germplasm. International Plant Genetic Resources Institute (IPGRI), Rome, Italy coordinates at global level. Various International Crop Research Institutes deal with germplasm of concerned crops. In India, National Bureau of Plant Genetic Resources (NBPGR). New Delhi deals with various aspects of germplasm of agricultural and horticultural crops. Forest Research Institute, Dehradun deals with germplasm of forest species and Botanical Survey of India, Kolkata deals wiith germplasm of remaining plant species. The role of IPGRI and NBPGR is briefly discussed as follows :

1. **IPGRI (OLD IBPGR):** International Plant Genetic Resources Institute (IPGRI) is an autonomous international scientific organization. The functioning of IPGRI and other International Research Institutes is supported and supervised by the Consultative Group on International Agricultural Research (CGIAR). CGIAR was constituted in 1971 by the joint efforts of Food and Agriculture Organization (FAO), the World Bank and United Nations Development Program (UNDP) to establish International Research Institutes and assess their progress. IPGRI was established by CGIAR in 1994 and it is situated in Rome, Italy at the Food and Agriculture Organization of the United Nations. The main function

of IPGRI is to conduct research and to promote an International Network of Plant genetic resources activities to ensure the collection, conservation, evaluation, documentation and utilization of plant germplasm. IPGRI also helps in global collection and exchange of plant genetic resources. IPGRI has constituted crop germplasm advisory committees to help in collection, conservation, evaluation, documentation and utilization of crop germplasm. Thus IPGRI promotes and coordinates the global collection and conservation programmes of plant genetic resources. Its predecessor till December 1993 was International Board for Plant Genetic Resources (IBPGR) which was established in 1974.

2. **NBPGR:** National Bureau of Plant Genetic Resources was established by Indian Council of Agricultural Research (ICAR) in 1976 in New Delhi. In India, plant introduction started in 1946 at IARI, New Delhi in the Division of Botany. In 1961 a separate division of Plant Introduction was established under the leadership of Dr. H.B. Singh. who made remarkable contribution in the field of Plant Introduction in India. He made huge collections of germplasm of various crop species and systematized the work. In 1976, the division of Plant Introduction was elevated to the status of independent agency known as NBPGR.

   The basic function of NBPGR is to conduct research and promote collection, conservation, evaluation, documentation and utilization of crop genetic resources in India. NBPGR is assisted by various Crop Research Institutes in the collection, conservation, evaluation and documentation of crop genetic resources. The main functions of NBPGR are briefly presented below:

   (i) NBPGR is the sole agency in India for import and export of plant genetic resources. Thus it helps in exchange of germplasm.

   (ii) It promotes national genetic resources activities, *viz.* collection, conservation, evaluation, documentation and utilization of crop plants, and coordinates in all these activities.

   (iii) NBPGR has five stations which are located at (1) Shimla, Himachal Pradesh, (2) Jodhpur, Rajasthan (3) Akola, Maharashtra (4) Kanya Kumari, Kerala, and (5) Shillong, Meghalaya. Collections of various crops are evaluated by these centres.

   (iv) NBPGR also organizes short term training courses on collection, conservation, evaluation, documentation and utilization of crop genetic resources.

   (v) National and International exploration and collection trips are also organized by NBPGR. National collection trips are organized in collaboration with the concerned Crop Research Institutes and International trips are arranged with the help of IPGRI/FAO.

   (vi) NBPGR provides guidance about development of cold storage facilities for medium and short term conservation of germplasm.

(vii) NBPGR also takes decision about setting up of gene sanctuaries for endangered crop species.

Q14. Explain briefly plant quarantine related to germplasnm.

Ans. There are chances of the entry of new diseases, insects and weeds when the material is collected and introduced from other countries. Several new weeds, diseases and insects have entered in India in the past alongwith plant collections. To prevent entry of new diseases, insects and weeds an Act was formulated in India in 1914 which is referred to as Destructive Insects and Pest Act.

Quarantine refers to a prophylactic means to prevent the entry of new diseases, insects and weeds from other countries. The quarantine offices are located at all seaports and international airports. Their main function is to examine the imported plant material for diseases, insects and weeds. They also check whether the material bears a Phytosanitary Certificate, which is issued by the sender and states that the material does not contain any disease, insect or weed. If the phytosanitary certificate is not available quarantine department may reject the material and destroy the same.

Three organizations look after the work of quarantine inspections in India for different types of plant materials. These organizations are: NBPGR, Forest Research Institute (FRI), Dehradun and Botanical Survey of India (BSI), Calcutta (Kolkata). NBPGR looks after the quarantine inspection work of the prop-gules of agricultural and horticultural crops. FRI deals with the quarantine inspection of forest trees prop-gules. BSI deals with quarantine inspection of the prop-gules of remaining plant species. Directorate of Plant Protection, Quarantine and Storage, Faridabad (Haryana) deals with the quarantine inspection of food grains and other produces imported for human consumption

Chapter 4

Genetics of Qualitative and Quantitative Characters

Q1. What are types of plant characters?

Ans. Plant characters are of two types, *viz.*, qualitative and quantitative. Qualitative traits are governed by one or two genes, whereas quantitative traits are governed by several genes. Those characters which are governed by several genes are called quantitative characters and variation observed in such characters is known as polygenic variation. Quantitative characters are also known as polygenic traits. The inheritance of traits governed by several genes is termed as polygenic inheritance. Examples of qualitative characters are color of stem, leaf, flower, seed, pollen; leaf shape and hairiness, *etc.* Quantitative or polygenic traits include yield, days to flower, days to maturity, seed size, seed oil content, protein content, seed weight, plant height, *etc.*

Q2. Explain briefly main features of quantitative traits.

Ans. There are some characteristic features of quantitative or polygenic characters. The important features of polygenic traits include (i) continuous variation, (ii) transgressive segregation, (iii) involvement of effective and ineffective alleles, (iv) environmental effect. These are briefly discussed as follows:

1. **Continuous Variation:** Polygenic traits exhibit continuous variation from one extreme to other. As a result clear cut classification of polygenic traits in different classes is not possible. In other words, in the segregation of polygenic traits, there is gradual variation.
2. **Transgressive Segregation:** Segregants which fall outside the range of both the parents are known as transgressive segregants. The appearance

of transgressive segregants in F_2 or later generation is an important feature of polygenic traits. Transgressive segregants are obtained when a cross is made between parents having intermediate values to the extreme value.

3. **Genes Involved:** Several genes are involved in polygenic variation. However, effect of individual gene is small which is not detectable. Two types of alleles, *viz.*, active and non-active are involved in the polygenic variation. These alleles are also known as contributing and noncontributing alleles or effective and non effective alleles. Those alleles which contribute to continuous variation are known as contributing alleles or active alleles or effective alleles. Those alleles which do not contribute to continuous variation are known as non-contributing alleles or inactive alleles or non-effective alleles.

4. **Environmental Effect:** Polygenic traits are highly sensitive to environmental changes. The environmental variation varies from 10 to 50 per cent for majority of polygenic traits. For some traits say yield, it is even more. The high environmental variation results in continuous variation. As a result effect of individual gene is not detectable.

Q3. What are several differences between polygenic and oligogenic characters?

Ans. There are several differences between polygenic and oligogenic characters [Table 4-1].

TABLE 4.1. Differences between Quantitative and Qualitative Traits

Sl.No.	*Particulars*	*Polygenic Traits*	*Oligogenic Traits*
1	Variation	Continuous	Discontinuous
2	Governed by	Several genes	One or few genes
3	Effect of individual gene	Not detectable	Detectable
4	Separation in to distinct classes	Not possible	Possible
5	Environmental effect	High	Little
6	Genes involved	Usually additive	Usually non additive
7	Analysis is based on	Mean, variances and co-variances	Frequency and ratios

Q4. What are similarities between qualitative and quantitative traits?

Ans. Quantitative characters do not differ from qualitative characters in any essential features (Mather, 1949; Falconer, 1960). The similarities between quantitative and qualitative traits are presented as follows:

(i) Both are governed by genes. The former is governed by minor genes and latter by major genes.

(ii) Genes for both types of traits are located on the chromosomes.

(iii) Both types of traits control variation. Polygenic traits control continuous variation, whereas oligogenic traits control discontinuous variation.

(iv) Both types of traits exhibit linkage, segregation, recombination, reciprocal differences and mutation.

(v) Expression of both types of traits is influenced by environmental factors. Polygenic traits are more sensitive to environmental factors than oligogenic traits.

(vi) Both types of traits exhibit gene interaction.

(vii) Polygenic traits exhibit transgressive segregation which can only be explained with the help of Mendelian principles of heredity.

Q5. Explain briefly the multiple factor hypothesis.

Ans. Multiple factor hypothesis was proposed by Nilsson Ehle in 1908 working with inheritance of kernel color in wheat. This hypothesis states that the inheritance of some plant characters is governed by several or multiple genes. Main features of multiple factor hypothesis are presented here as under.

(i) This is also known as theory of polygenic traits.

(ii) This hypothesis provides explanation to the inheritance of quantitative traits which exhibit continuous variation from one extreme to another for a character.

(iii) In polygenic inheritance, effect of each gene is not easily detectable.

(iv) For polygenic traits, classification of plants into distinct clear cut classes is not possible. Analysis of polygenic characters is based on means, variances and co-variances.

(v) Polygenic traits exhibit transgressive segregation.

(vi) Multiple factor hypothesis is universally accepted for explanation of polygenic traits.

Q6. What are the components of polygenic variation?

Ans. The polygenic variation present in the plant breeding populations is of three types, *viz.*, phenotypic, genotypic and environmental. These are explained as follows:

1. **Phenotypic Variation:** It is the total variation which is observable. It includes both genotypic and environmental components and hence changes under different environmental conditions. Such variation is measured in terms of phenotypic variance.

2. **Genotypic Variation:** It is the inherent variation which remains unchanged by environmental factors. This type of variation is more useful to a plant breeder for exploitation ib selection or hybridization. Such variation is measured in terms of genotypic variance. The genotypic variation consists of additive, dominance and environmental components.

3. **Environmental variation:** It refers to non-heritable variation which is entirely due to environmental effects and varies under different environmental conditions. This is uncontrolled variation which is measured in terms of error mean variance. The variation in true breeding parental lines and their F1 hybrid is non-heritable.

Q7. What is the significance of polygenes?

Ans. Polygenes are of prime importance to the plant breeder for evolution of improved cultivars (Mather and Jinks, 1971; Simmonds, 1979). Polygenes have great evolutionary significance. They provide variation for fine adjustment and are systems of smooth adaptive change and speciation. It is believed that in natural populations, the best adapted or the most fit individuals are those that are close to the population mean (have intermediate value) for various quantitative traits.

Q8. What are the types of polygenic variability?

Ans. Mather (1943) has recognized two types of polygenic variability, *viz.*, free variability and potential variability. He has nicely explained the mechanism of storage and release of polygenic variability. These are discussed as follows:

1. **Free Variability:** It refers to phenotypic differences between homozygotes with extreme phenotypes (AABBCC and aabbcc). Such variability is expressed and exposed to selection. Natural selection acts against extreme phenotypes and favors intermediate types.
2. **Potential Variability:** It refers to hidden or bound variability in the heterozygotes or in the homozygotes which do not have the extreme phenotype and, therefore, is not exposed to selection. The potential variability is released after crossing of such genotypes with other genotypes followed by segregation and recombination. It is of two types, *viz.*, heterozygotic and homozygotic. These are explained as follows:
   (i) **Heterozygotic Potential Variability:** This type of variability is stored in heterozygote *e.g.*, AaBb. Such heterozygotes are phenotypically uniform and are very close to the population mean. However, they would produce extreme phenotypes in the next generation due to segregation and recombination. Thus, the heterozygotes function as stores of variability which is released slowly as free variability due to segregation and recombination.
   (ii) **Homozygotic Potential Variability:** Homozygotes also function as stores of variability. For example, Two gene homozygotes AAbb and aaBB may be expected to cluster around the mean of the population. They would, therefore, be protected from natural selection and would be phenotypically uniform. However, they would produce the extreme phenotypes AABB and aabb after intermating *i.e.* AAbb x aaBB followed by segregation and recombination. The release of this type of variability is slow because it must first be converted into

heterozygotic potential variability through hybridization and then it is released as free variability.

The above concepts have been explained based on two genes. In case of polygenic traits several genes are involved in the expression of a character and there is linkage between genes having plus and minus effects on a trait. The linkage among polygenes is useful. It reduces immediate resposnse to selection but prolongs the response to selection due to slow release of potential genetic variability in the segregating generation.

Section II

Breeding Methods

Chapter 5

Genetic Basis of Breeding Methods

Q1. What are main features of self-pollinated species?

Ans. These are self-fertilizing species. In these species, development of seed takes place by self-pollination (autogamy). Hence, self-pollinated species are also known as autogamous species or inbreeders. Various plant characters such as homogamy, cleistogamy, chasmogamy, bisexuality *etc.* favor self- fertilization. Some important features of autogamous species are given below:

(i) They have regular self-pollination.

(ii) They are homozygous and have advantage of homozygosity, means they are true breeding.

(iii) In-breeders do not have recessive deleterious genes, because deleterious genes are eliminated due to inbreeding by way of gene fixation.

(iv) In-breeders have homozygous balance and, therefore, are tolerant to inbreeding. In other words, inbreeding does not have any adverse effects on in-breeders.'

(v) In autogamous species, new gene combination are not possible due to regular self-pollination. 6. Inbreeders are composed of several component (homozygous) lines. Hence, variability is mostly among component lines.

(vi) Inbreeders have generally narrow adaptation and are less flexible.

Q2. What are main features of cross-pollinated species?

Ans. This group refers to cross fertilizing species. These species produce seed by cross pollination (allogamy). Hence, these species are also referred to as allogamous species or out-breeders. Various plant characters which promote cross pollination include dichogamy, moecy, dioecy, heterostyly, herkogamy, self-incompatibility and male sterility. Some important features of outbreeders are given as follow:

(i) They have random mating. In such population, each genotype has equal chance of mating with all other genotypes.

(ii) Individuals are heterozygous and have advantage of heterozygosity.

(iii) Individuals have deleterious recessive genes which are concealed by masking effect of dominant genes.

(iv) Out-breeders are intolerant to inbreeding. They exhibit high degree of inbreeding depression on selfing.

(v) Cross pollination permits new gene combinations from different sources.

(vi) In these species, variability is distributed over entire population.

(vii) They have wide adaptability and more flexibility to environmental changes due to heterozygosity and heterogeneity.

Q3. What are asexually propagated species?

Ans. Some crop plants propagate by asexual means *i.e.* by stem or root cuttings or by other means. Such species are known as asexually propagated species or vegetatively propagated species. Such species are found in both self and cross pollinated groups. Generally, asexually propagated species are highly heterozygous and have broad genetic base, wide adaptability and more flexibility.

Q4. What are genetic constitutions of breeding populations?

Ans. The genetic constitution of plants is determined by mode of pollination. Self-pollination leads to homozygosity and cross pollination results in heterozygosity. Thus, we have to exploit homozygosity in self-pollinated crops and heterozygosity in cross pollinated species, because in-breeders have advantage of homozygosity and out-breeders have advantage of heterozygosity. Based on genetic constitution, plant breeding populations are of four types, *viz.* (1) homogeneous, (2) heterogeneous, (3) homozygous, and (4) heterozygous. These are briefly discussed below:

1. **Homogeneous Populations:** A population with genetically similar plants is called homogeneous populations. Main features of such population are as follows:

   (i) All individuals of such population have exactly the same genetic constitution.

   (ii) Such population breeds true on self- pollination.

   (iii) Such population has narrow genetic base

   (iv) There is no genetic variation in such population.

   (v) Examples of such populations are pure lines, inbred lines, F_1 hybrid between two pure lines or inbred lines and progeny of a clone.

2. **Heterogeneous Populations:** A population with genetically dissimilar plants constitute is called heterogeneous populations. Main features of such population are as follows:

   (i) Such population has genetic variation.

   (ii) Such population is a mixture of several homozygous lines.

   (iii) Such population breeds true on self- pollination.

   (iv) Such population has broad genetic base

   (v) Examples of such populations are land races, mass selected populations, composites, synthetics and multi-lines.

   (vi) Heterogeneous populations have wide adaptability and stable performance under different environments.

Table 5-1. Comparison of Homozygous and Heterozygous Populations

Sl.No.	Particulars	Homogeneous Population	Heterogeneous Population
1	Genetic constitution of plants	Similar	Dissimilar
2	Progeny of	Single homozygote	Mixture of several homozygotes
3	Genetic variation	Absent	Present
4	Effect of selfing	Breed true (except F1)	Breed true
5	Adaptation	Narrow	Wide or broad
6	Examples	Pure lines, Inbred lines and F1	Mass selected variety, composites and synthetics

3. **Homozygous Populations:** A population of individuals with like-alleles at the corresponding loci is known as homozygous population. Main features of such population are presented as follows:

   (i) Such population do not segregate on self-pollination. Thus non-segregating genotypes constitute homozygous populations.

   (ii) Such population may be of two types, *viz.* homozygous homogeneous and homozygous heterogeneous.

   (iii) Example of homozygous homogeneous populations are pure lines, and inbred lines.

   (iv) Examples of homozygous and heterogeneous populations are mass selected populations multiline in self-pollinated plants.

   (v) Pure lines and inbred lined do not have heritable variation.

   (vi) Mass selected cultivars have heritable variation

4. **Heterozygous Populations: A population of** individuals with unlike-alleles at the corresponding loci is referred to as heterozygous population. Main features of such populations are as follows:

   (i) Individuals of such population segregate into various types on selfing.

(ii) Such populations have greater buffering capacity to environmental fluctuations.

(iii) Such population is of two types, *viz.* heterozygous and heterogeneous and heterozygous and homogeneous.

(iv) Examples of first type include composites and synthetics.

(v) Example of second type include F_1 hybrids between two pure lines or inbred lines and clonal variety,

Q5. Give Classification of plant breeding methods based on application.

Ans. The choice of breeding methods mainly depends on the mode of pollination, mode of reproduction, gene action and breeding objective of crop species. Plant breeding methods are generally classified on the basis of application in crop improvement (general methods, special methods and population improvement approaches) and hybridization (methods involving hybridization and methods not involving hybridization). Based on application plant breeding methods are of three types, *viz.* general breeding methods, special breeding methods and population improvement approaches.

1. **General Breeding Methods**: Various breeding procedures that are more commonly used for the genetic improvement of various crop plants are known as general breeding methods. Such breeding methods are of following three types: Such methods include Plant Introduction, Pure line selection, mass selection, progeny selection, pedigree method, bulk method, back cross method, SSD, clonal selection, heterosis breeding, synthetics and composites.
2. **Special Breeding Methods:** Those breeding procedures that are rarely used for improvement of crop plants are referred to as special breeding methods. Such methods include: mutation breeding, polyploidy breeding, wide crossing or distant hybridization, transgenic breeding, molecular breeding and somatic hybridization.
3. **Population Improvement Approaches:** Those breeding approaches which are used for accumulating favorable gene in a population are called population improvement approaches. Such approaches include: Recurrent selection, disruptive selection, diallel selective mating system and biparental mating.

Q6. Give Classification of plant breeding methods based on hybridization.

1. **Methods Involving Hybridization:** Such breeding methods include, Pedigree, bulk, backcross and SSD Methods; heterosis breeding and development of synthetics, composites. And population improvement approaches.
2. **Methods Not Involving Hybridization:** Such breeding methods include: Plant Introduction, pure line selection, mass selection, progeny selection, clonal selection, mutation breeding and transgenic breeding.

Q7. What are the differences in the breeding methods used for self-pollinated and cross pollinated species?

Ans. There are some differences in the breeding methods used for self-pollinated and cross pollinated species. Some important points are presented as follows:

1. Self-pollinated species are homozygous, hence we can start hybridization directly.
2. Cross pollinated species, on the other hand, are highly heterozygous. Hence we cannot start hybridization directly. First we have to develop inbred lines by selfing or inbreeding and then only hybridization can be taken up.
3. We have to exploit homozygosity in self-pollinated crops and heterozygosity in cross pollinated species.
4. Asexually propagated species such as sugarcane, potato, sweet potato, *etc.*, are highly heterozygous. Hence, F1 hybrids in such crops exhibit segregation and selection can be practiced in F1 generation.
5. The superior clones are identified and further multiplied. The maintenance or conservation of hybrid vigor is easy in such crops because of asexual propagation.

Q8. What are methods of breeding autogamous species?

Ans. Plant breeding methods that are used for genetic improvement of self-pollinated or autogamous species include: (1) Plant Introduction, (2) Pureline selection, (3) Mass selection, (4) Pedigree method, (5) Bulk method, (6) Single seed descent method, (7) Backcross method, (8) Heterosis breeding, (9) Mutation breeding, (10) Polyploidy breeding, (11) Distant hybridization and (12) Transgenic breeding. Four breeding approaches, *viz.* recurrent selection, disruptive selection, diallel selective mating, and biparental mating are used for population improvement.

Q9. What are methods of breeding allogamous species?

Ans. Breeding methods that are used for genetic improvement of cross pollinated or allogamous species include: (1) Plant Introduction, (2) Mass and progeny selection, (3) Backcross method, (4) Heterosis breeding, (5) Synthetic breeding, (6) Composite breeding, (7) Polyploidy breeding, (8) Distant hybridization, and (9) Transgenic breeding. Mutation breeding is rarely used in allogamous species. Three breeding approaches, *viz.*, recurrent selection, disruptive mating and biparental mating are used for population improvement.

Q10. What are methods of breeding asexually propagated species?

Ans. Important breeding methods applicable to asexually propagated species are: (1) Plant Introduction, (2) clonal selection, (3) Mass selection, (4) Heteross breeding, (5) Mutation breeding, (6) Polyploidy breeding, (7)

Distant hybridization, and (8) Transgenic breeding. Mass selection is rarely used in asexually propagated species.

Q11. Explain genetic basis of varieties developed by various breeding methods.

Ans. A brief account of the genetic basis of varieties developed by various breeding methods is presented as follows:

1. **Introduced Variety:** Plant introduction is applicable to all three groups of crop plants, *viz.*, self-pollinated, cross pollinated and asexually propagated species. It is an oldest and rapid method of crop improvement. The genetic base of an introduced variety depends on its method of development and mode of pollination. It may be a homogeneous population, heterogeneous population and hetezygous population depending upon method of its development.
2. **Pure line Variety:** Pure line selection is applicable to self-pollinated species. It is also used sometimes in cross pollinated species for development of inbred lines. A single best pure line is released as a variety. Thus a pure line variety is homozygous and homogeneous population.
3. **Mass Selected Variety:** Mass selection is common in cross pollinated species and rare in self-pollinated and asexually propagated species. In self-pollinated crops, a mass selected variety is a mixture of several pure lines. Thus it is a homozygous but heterogeneous population. In cross pollinated species, a mass selected variety is a mixture of several hetero and homozygotes. Thus, it is a heterozygous and heterogeneous population.
4. **Progeny Selected Variety:** Progeny selection is used in cross pollinated species. A variety developed by this method is heterozygous and heterogeneous population because it consists of several hetero and homozygotes.
5. **Pedigree Bred Variety:** Pedigree method is applicable to both self and cross pollinated species. In self-pollinated crops progeny of a single best homozygote is released as a variety. Thus a variety developed by this method has a homozygous and homogeneous population. In cross pollinated species, it is used for development of inbred lines.
6. **Bulk Breeding and SSD:** Bulk and single seed descent methods are used in self-pollinated species. Progeny of a single best homozygote is released as a variety by these methods. Thus, varieties developed by these methods are homozygous and homogeneous.
7. **Backcross Bred Variety:** Backcross method is applicable in all three groups of crop species. This method is used for transfer of oligogenic characters from a donor source to a well adapted variety. This method is also used for development of multi-lines, isogenic lines and transfer of male sterility. This method is more effective in transferring oligogenic characters than

polygenic traits. The end product of backcross method is similar to parent variety except for the character which has to be transferred from the donor source. It is a progeny of best single homozygote. In self-pollinated species, varieties developed by back cross method are homozygous and homogeneous.

8. **Multi-line Variety:** Multiline varieties are developed in self-pollinated species. They are mixture of several isogenic lines, closely related lines or unrelated lines. Thus, a multiline variety is a homozygous but heterogeneous population.

9. **Clonal variety:** Clonal selection is used in asexually propagated species. In this method progeny of a single best clone is released as a variety. Such variety has heterozygous but homogeneous population.

10. **Hybrid Variety:** Heterosis breeding is used in all the three groups. However, it is common in cross pollinated and asexually propagated species and rare in self-pollinated species. A hybrid variety has homogeneous but heterozygous population.

11. **Synthetic and Composite Varieties:** Synthetic and composite varieties are developed in cross pollinated species. Such varieties consist of several homozygotes and heterozygotes and thus constitute a heterogeneous population.

12. **Mutant Variety:** Mutation breeding is common in self-pollinated and asexually propagated species and rare in cross pollinated species. A mutant variety differs from parent variety in one or few characters. A Mutant differs from a segregant in two main ways. Firstly, the frequency of segregants is very high and that of mutant is extremely low (0.1 per cent). Secondly, mutant differs from parent variety in one or few characters, where as a segregant differs from parent material in several characters. Polyploidy breeding is common in asexually propagated species and rare in self and cross pollinated species. A polyploid variety differs from parent variety in chromosome numbers and exhibit gigant morphological characters.

13. **Product of Distant Hybridization:** Distant hybridization is used in all the three types of crop species. However, this method is used for transferring some desirable genes from wild species to the cultivated ones. Generally, backcross method is used for transfer of oligogenic characters and pedigree method for transfer of polygenic characters.

14. **Transgenic Variety:** Transgenic breeding is applicable to all three types of crop species. This method is used to solve specific problems which cannot be solved by conventional breeding techniques. This method will serve as a tool and cannot be used as a substitute for conventional breeding methods. Transgenic variety is similar to parent variety except for the gene incorporated.

15. **End Product of Recurrent Selection:** Recurrent selection is common in cross pollinated species and rare in other two groups. It is used for accumulating favorable genes in a population *i.e.* for population improvement. Other approaches which are used for population improvement include disruptive mating, diallel selective mating (DSM) and biparental mating. DSM is used in self-pollinated species and other two techniques can be used both in self and cross pollinated species.

Chapter 6

Breeding Self Pollinated Crops (Plant Introduction and Pure Line Selection)

Q1. What are breeding methods?

Ans. Various approaches (*viz.* selection, hybridization, mutation, *etc.*) that are used for genetic improvement of crop plants are referred to as plant breeding methods or plant breeding procedures or plant breeding techniques. The choice of breeding methods mainly depends on four things *viz.*, mode of pollination and reproduction, gene action and breeding objective of crop species. Plant breeding methods are generally classified on the basis of application in crop improvement (general methods, special methods and population improvement approaches) and hybridization (methods involving hybridization and methods not involving hybridization).

Q2. Give a classification of breeding methods based on application.

Ans. Based on application, breeding methods are of three types, *viz.* general breeding methods, special breeding methods and population improvement approaches. These are explained as follows:

1. **General Breeding Methods**: Various breeding procedures that are more commonly used for the genetic improvement of various crop plants are known as general breeding methods. Such methods include Plant Introduction, Pure line selection, mass selection, progeny selection, pedigree method, bulk method, back cross method, SSD, clonal selection, heterosis breeding, synthetics and composites.

2. **Special Breeding Methods:** Those breeding procedures that are rarely used for improvement of crop plants are referred to as special breeding methods. Such methods include: mutation breeding, polyploidy breeding, wide crossing or distant hybridization, transgenic breeding, molecular breeding and somatic hybridization.
3. **Population Improvement Approaches:** Those breeding approaches which are used for accumulating favorable gene in a population are called population improvement approaches. Such approaches include: Recurrent selection, disruptive selection, diallel selective mating system and biparental mating.

Q3. Give a classification of breeding methods based on hybridization.

Ans. Based on hybridization, breeding methods are of two types, *viz.* methods involving hybridization and methods not involving hybridization. These are explained as follows:

1. **Methods Involving Hybridization**: Such breeding methods include: Pedigree, bulk, backcross and SSD Methods; heterosis breeding and development of synthetics and composites and population improvement approaches.
2. **Methods Not Involving Hybridization:** Such breeding methods include: Plant Introduction, pure line selection, mass selection, progeny selection, clonal selection, mutation breeding and transgenic breeding.

Q4. What are differences in the breeding methods used for self-pollinated and cross pollinated species?

Ans. There are some differences in the breeding methods used for self-pollinated and cross pollinated species. Some important points are presented as follows:

1. Self-pollinated species are homozygous, hence we can start hybridization directly.
2. Cross pollinated species, on the other hand, are highly heterozygous. Hence we cannot start hybridization directly. First we have to develop inbred lines by selfing or inbreeding and then only hybridization can be taken up.
3. We have to exploit homozygosity in self-pollinated crops and heterozygosity in cross pollinated species.
4. Asexually propagated species such as sugarcane, potato, sweet potato, *etc.*, are highly heterozygous. Hence, F_1 hybrids in such crops exhibit segregation and selection can be practiced in F_1 generation.
5. The superior clones are identified and further multiplied. The maintenance or conservation of hybrid vigor is easy in such crops because of asexual propagation.

Q5. What are the methods of breeding self-pollinated crops?

Ans. Plant breeding methods that are used for genetic improvement of self-pollinated or autogamous species include: (1) Plant Introduction, (2) Pure line selection, (3) Mass selection, (4) Pedigree method, (5) Bulk method, (6) Single seed descent method, (7) Backcross method, (8) Heterosis breeding, (9) Mutation breeding, (10) Polyploidy breeding, (11) Distant hybridization and (12) Transgenic breeding. Four breeding approaches, *viz.* recurrent selection, disruptive selection, diallel selective mating, and biparental mating are used for population improvement.

Q6. What are the methods of breeding cross-pollinated crops?

Ans. Breeding methods that are used for genetic improvement of cross pollinated or allogamous species include: (1) Plant Introduction, (2) Mass and progeny selection, (3) Backcross method, (4) Heterosis breeding, (5) Synthetic breeding, (6) Composite breeding, (7) Polyploidy breeding, (8) Distant hybridization, and (9) Transgenic breeding. Mutation breeding is rarely used in allogamous species. Three breeding approaches, *viz.*, recurrent selection, disruptive mating and biparental mating are used for population improvement.

Q7. What are methods of breeding asexually propagated species?

Ans. Important breeding methods applicable to asexually propagated species are: (1) Plant Introduction, (2) clonal selection, (3) Mass selection, (4) Heteross breeding, (5) Mutation breeding, (6) Polyploidy breeding, (7) Distant hybridization, and (8) Transgenic breeding. Mass selection is rarely used in asexually propagated species.

Non hybridization methods commonly used in the genetic improvement of self pollinated crops include introduction and pure line selection. This chapter deals with these two methods of crop improvement.

Q8. What is plant introduction? Give its main features.

Ans. The transposition of crop plants from the place of their cultivation to such areas where they were never grown earlier is called plant introduction. Plant introduction is the process of taking crop plants into new areas where they were never cultivated earlier.

Plant introduction is an ancient method of crop improvement. The main points related to plant introduction are briefly discussed below:

(i) Climatic Difference: The place from where the variety or plant material is taken and the place where the material is introduced differ in agro-climatic conditions. Plant introduction usually is done from one country to another. But sometimes it can take place between two climatic regions of the same country.

(ii) Methods of Introduction: Earlier, the introduction of plants from one place to another used to be done by travelers, traders, and merchants.

Now special organizations have been established for this work. IPGRI coordinates the work of plant introduction on global basis.

(iii) Methods of Collection: The plant material is obtained through correspondence, personal visit, exchange of material, purchase or as gift.

(iv) Plant Material to be Introduced: The material of seed propagated crops is introduced in the form of seed and that of vegetatively propagated crops in the form of cuttings or propagules.

(v) Plant Introduction Organizations: In India plant introduction work for agricultural crops and horticultural plants is carried out by the National Bureau of Plant Genetic Resources[NBPGR], New Delhi; for forest plants by Forest Research Institute [FRI], Dehradun; and for remaining plants by Botanical Survey of India [BSI], Kolkata.

(vi) Crop Plants Introduced: Soon after the discovery of New World many crops plants were introduced from one part of the World to another as given below:

(a) From New World to Old World: The crops which were introduced to the Old World included maize, beans, peanuts, potato, tomato, tobacco, American cotton, rubber, peppers and squash. (b) From Old World to New World: The crops which were brought to the New World included wheat, barley, oats, rice, sugarcane, bananas, coffee, many fruits, vegetable* and forage crops.

(vii) Application: Plant introduction method is applicable for improvement of all three groups of plants, *viz.*, self pollinated, cross pollinated and vegetatively propagated crops.

Q9. What are types of plant introduction?

Ans. Plant introductions are generally classified on the basis of adaptation and utilization. Base on adaptation, plant introductions are of two types, *viz.* (1) primary introduction, and (2) secondary introduction. Based on utilization, again introductions are of two types, *viz.* (i) direct introduction, and (ii) indirect introduction. These are defined as follows:

1. **Primary Introduction:** Introductions that can be used for commercial cultivation as a variety without any change in the original genotype is called primary introduction. Examples of primary introductions are Sonora 64 and Lerma Rojo in wheat and Taichung Native 1, IR 8, IR 20 and IR 36 in rice. Introduction that are immediately adapted to the changed environment are known as direct introductions. Thus primary introductions can also be called as direct introductions. Any foreign variety which is directly recommended for commercial cultivation in the new environment is called exotic variety.

2. **Secondary Introduction:** Introduction that can be used as a variety after selection from the original variety or used for transfer of some desirable

gene to the cultivated variety is known as secondary introduction. Examples of secondary introductions are Kalyan Sona and Sonalika varieties of wheat which were released after selection from the material received from Mexico.

Q10. What are uses of plant introductions?

Ans. Plant introductions are utilized in crop improvement in three main ways: *viz*,, (i) direct as a variety, (ii) as a variety after selection, and (iii) as a parent in the hybridization for the development of improved variety.

1. **As a variety:** In some crops, the introduced material is directly released as a new variety. Some examples of direct release of introduced material in India are given below:

   (a) **Wheat**: Semi-dwarf varieties Sonora 64 and Lerma Rojo.

   (b) **Paddy**: Semi-dwarf varieties Tajchung Native 1, IR 8, IR 20 and IR. 36.

   (c) **Soybean**: Varieties Bragg and Lee.

   There are many more examples where introduced material was found of direct use and released as new variety for commercial cultivation.

2. **As a new variety after selection:** Sometimes, the introduced material is not found useful as such. In such case efforts are made to develop new varieties through selection. There are several crops in which new varieties have been developed through selection from introductions. Some examples are as follows:

   (a) **Egyptian Cotton**: The variety Sujata was released after selection from the Egyptian variety Kamak,

   (b) **American Cotton**: The variety PRS 72 was released after selection from Russian material.

   (c) **Wheat**: In wheat, varieties Kajyan Sona and Sonalika are the result of selection from Mexican wheat introductions.

   Similarly, new varieties have been developed through selection from the introductions in pearl millet (Improved Ghana), cowpea, radish, sweet potato and many other crops.

3. **As a parent in hybridization:** Introductions are widely used as parents in the hybridization programmes for the development of new varieties in almost all important agricultural and horticultural crops. For example, all semi-dwarf varieties of wheat and paddy have been developed through the use of introduced material. Many varieties and hybrids in maize, sorghum, and pearl millet have been developed involving introduced material as one of the parents. The pioneer cotton hybrid *H4* has been developed from a cross between Gujarat 67 x American nectariless. Here the male parent is an introduction from America.

Q11. What are merits and demerits of plant introduction?

Ans. Plant Introduction as a method of crop improvement has several merits and demerits which are briefly presented below:

Merits

(i) It is a very easy and quick method of developing new varieties especially when introduced material is used directly as a variety or after selection.

(ii) This method is useful in introducing new crop plants. For example, crops like maize, potato, tomato, groundnut, papaya, pine apple, Triticale, *etc.* were introduced in India from other countries.

(iii) This is a good method of collection and conservation of germplasm of different crops to protect the same from genetic erosion.

(iv) This is an effective method of conserving those crop species which have been threatened by the danger of extinction. Such species can be saved by shifting them to other areas.

(v) This method is applicable in all self pollinated, cross pollinated and vegetatively propagated crops.

Demerits

The main demerit of this method is that there are chances of the entry of new diseases, insects and weeds in the country along with the introduced material. For example, *Argemone mexicana* weed, late Blight of potato, bunchy top of banana, and coffee rust diseases, and woolly aphids of apple and potato tuber moth insects were introduced in India along with introduced material. However, all these entered before the establishment of quarantine organization. Now chances of entry of new insects, diseases and weeds are remote due to quarantine regulations and checkup.

Q12. What is selection?

Ans. Selection refers to the process that favours survival and further propagation of some plants having more desirable characters than others. The end product of selection process is also known as selection. Selection is of two types, *viz.* natural selection and artificial selection. Natural selection operates in nature without human interference. Natural selection favours those characteristics that are essential for survival of a species. Thus survival or adaptation is the main concern in the natural selection. Artificial selection, on the other hand is made by human. It favours those characteristics of plants that are related to yield and quality.

Q13. What is pure line selection? Give its main features.

Ans. Pure line refers to the homogeneous progeny of a self- pollinated homozygous plant. Development of new variety through identification and isolation of single best plant progeny is known as pure line selection or individual plant selection. This method is commonly used in self pollinated species. The main features of pure lines are briefly presented below:

(i) Pure lines are homozygous and homogeneous.

(ii) The variation within a pure line is entirely due to environmental factors. Thus the variation is non- heritable in the pure lines.

(iii) A variety developed by pure line selection is highly uniform in quality due to absence of genetic variation.

(iv) Selection is ineffective in a pure line due to lack of heritable variation. Selection is effective when heritable variation is present.

(v) Generally, pure line varieties have narrow adaptation and poor adaptability than heterogeneous populations. The poor adaptability is due to narrow genetic base.

(vi) Pure line varieties are more prone to the attack of new diseases due to genetic uniformity and narrow genetic base.

(vii) Pure lines can be isolated from heterogeneous population as well as segregating populations through individual plant selection and progeny testing.

(viii) In a pure line variety, natural outcrossing, mutations and mechanical mixtures are the important sources of genetic variation. The spontaneous mutations can not be controlled. Other two factors can be controlled.

Johannsen (1903, 1926), a Danish biologist developed the concept of pure line theory working with Princess variety of common bean (*Phaseolus vulgaris*). He concluded that:

(i) Continuous inbreeding (selfing) leads to homozygosity,

(ii) Variation within a pure line results from environmental factors only,

(iii) Selection within a pure line is not effective because all the plants in a pure line have exactly the same genotype, and

(iv) Selection in the original population is effective because the plants have genetic variation.

Q14. Explain briefly the procedure of pure line selection.

Ans. Pure line selection consists of four major steps, *viz.* (i) selection of a heterogeneous population from which pure lines have to be isolated, (ii) Isolation of pure lines by individual plant selection, (iii) testing of pure lines in field trials, and (iv) release of the best pure line as a variety. The year wise procedure is given as follows:

1. **First year.** An old variety or land race is used for pure line selection. Single plants are selected from the heterogeneous population keeping in view the objective of selection. The number of individual plants to be selected may vary from 200 to 1000 in various crops.

2. **Second year.** The progeny of each selected plant is grown separately in few rows and evaluated for the character under consideration The top 15-20 progenies are selected and seed of all plants in each progeny is bulked which constitutes strains.

3. **Third year.** The strains constituted in second year are evaluated in replicated field trials and top performing few strains are selected for further evaluation.
4. **Fourth to Seventh year.** The selected strains are evaluated in field trials for 2-3 years for yield performance. In India, the selected entries (strains) are evaluated in All India Coordinated Crop Improvement Project. The best genotype is identified on the basis of yield performance.
5. **Eighth to tenth year.** The best performing strain is released and notified as a variety. Then the breeder, foundation and certified seeds are produced. The production of certified seed takes two years after release of a variety. Thus the seed of new variety reaches the farmers in tenth year.

Q15. What are the merits and demerits of pure line selection?

Ans. The important merits and demerits of pure line selection are briefly presented below:

Merits

(i) This is a good method of isolating the best genotype for yield, disease resistance, insect resistance, earliness, quality, *etc.* from a heterogeneous, or mixed population of an old variety.

(ii) The variety developed by this method is uniform and more attractive than mass selected variety.

(iii) This is an easy and cheap method of crop Improvement.

Demerits

(i) This method can isolate only superior genotypes from the mixed population. It cannot develop new genotypes.

(ii) This method is applicable to self pollinated species only. It cannot be used for development of variety in cross pollinated species.

(iii) The varieties developed by pure line selection have poor adaptability due to narrow genetic base. All the plants of a pure line have identical genotypes. Hence, such varieties are more prone to the attack of new diseases due to genetic uniformity.

Q16. What are the achievements of pure line selection?

Ans. Superior varieties have been isolated by pure line selection from the heterogeneous populations in several self pollinated crops like wheat, barley, paddy, sorghum, peanut, chickpea, pigeon pea, black gram, green gram, linseed, cowpea, *etc.*

Chapter 7

Breeding Self Pollinated Crops (Pedigree, Bulk, SSD and Backcross Methods)

Q1. What are breeding methods involving hybridization?

Ans. There are four hybridization methods which are commonly used for genetic improvement of self pollinated crops. These are pedigree breeding, bulk breeding, single seed descend and backcross breeding. Hybridization is used when there is limited genetic variability in a species and improvement is not possible by pure line selection and mass selection. Hybridization may involve genotypes of the same species, different species and different genera of the same family. Hybridization is defined as follows: Artificially crossing between genetically dissimilar plants is called hybridization.

Q2. What are types of hybridization?

Ans. Depending upon the involvement of genotypes, hybridization is of following three types:

1. **Intra-specific hybridization:** It involves two different genotypes of the same species. It is also called inter-varietal hybridization. This type of hybridization is widely used in plant breeding for development of varieties and hybrids.
2. **Interspecific hybridization:** It involves two different species belonging' to the same genus. It is also known as intrageneric hybridization. It is used for transfer of specific character from one species to another species. It is lesser common than inter-varietal hybridization.
3. **Inter-generic hybridization:** It involves two species each belonging to different genus of a family. It is rarely used for transfer of specific character

from genus to another. This type of hybrids are always sterile. The fertility is restored by doubling of chromosomes by colchicine treatment.

Inter-varietal hybridization is extensively used and other two types of hybridization are used for specific purposes.

Q3. What are objectives of hybridization?

Ans. Hybridization is carried out for various purposes. The important objectives of hybridization are presented as follows:

(i) To combine several desirable traits from different sources into a single genotype,

(ii) To create vast genetic variability for various economic characters in a population,

(iii) To develop superior crop cultivars, and

(iv) To develop hybrid varieties for commercial cultivation.

There are four methods of handling segregating populations (*viz.*, *F1*, F3, *F4*, ETC.) after hybridization. These are pedigree breeding, bulk breeding, single seed descend and backcross breeding.

Q4. Describe briefly main features of pedigree breeding method.

Ans. Pedigree refers to record of the ancestry of an individual selected plant. Pedigree breeding is a method of genetic improvement of self pollinated species in which superior genotypes are selected from segregating generations and proper record of the ancestry of selected plants are maintained in each generation. In other words, it is a selection procedure in segregating population of self pollinated species that keeps proper record of plants and or progeny selected in each generation. Main features of this breeding method are given below:

(i) Application. This method is widely used for the improvement of self pollinated species. It is generally used when both the parents that are used in the hybridization have good agronomic characters or are well adapted. Moreover, it is more commonly used for the improvement of polygenic traits than oligogenic characters.

(ii) Maintenance of pedigree records. In this method proper record of the ancestry of each selected plant or plant progeny is maintained for all generations of selection. Important characters of each selected plant and progeny are recorded.

(iii) Selection. In this method only human selection or artificial selection is used. Natural selection is allowed to operate only in the modified form of pedigree breeding known as mass pedigree method (*see* later).

(iv) Time taken. Development of new crop cultivar by this method generally takes 14-15 years.

(v) Genetic constitution. The variety developed by this method is homozygous and homogeneous, because it is a progeny of single homozygote.

Q5. Describe in brief breeding procedure of pedigree method.

Ans. First parents are selected keeping in view the breeding objective. The cross is made between selected parents. The *F1* material is grown using wide spacings. The dominance behavior for various characters is recorded. In *F2* also the material is grown using wide spacing. Individual plant selection is practiced in *F2*. The progeny of each selected plant is grown separately which constitutes *F3* generation. In *F3* and *F4* generations, selection is practiced within and between families. From *F5* to *F8* between progeny selection is practiced and superior progenies are identified and isolated in *F8*. These progenies constitute strains. These strains are evaluated in replicated multi-location trials for a period of 3-5 years. Based on superior performance, the strain is released as a variety (Figure 12.1). Thus release of new variety by this method takes 14-15 years. The number of plants be grown and selected in each generation is presented in Table 12.1. This is a general procedure of pedigree method. The number of plants to be grown and selected in each generation is not fixed. It may slightly vary from crop to crop.

TABLE 7-1. Procedure of Pedigree Breeding Method

Generation	*Year*	*No. of Plants/Progeny to be Planted*	*No. of Plants/Progeny to be Selected*
F1	1	75 plants	All 75 plants
F2	2	10,000 plants	1000 plants
F3	3	1000 plants	200 plants
F4	4	200 progeny	130 Plants
F5	5	130 progeny	50 progeny
F6	6	50 progeny	25 progeny
F7	7	25 progeny	10 progeny
F8	8	10 progeny	Preliminary yield trial.
	9-13	10 strains	Multi-location trial and identification of superior strain.
	14		Release of the best strain as a variety.

Q6. What is mass pedigree method?

Ans. It is a modification of pedigree method. It was proposed by Harrington in 1937. Mass pedigree method refers to growing of segregating material by bulk (mass) method when conditions are unfavorable for selection and use of progeny testing (Pedigree method) when conditions are favorable for selection. The bulk period may vary from one to few generations depending upon the occurrence of favorable conditions. The bulk period is-terminated as soon as favorable condition for selection occur. For example, in wheat dry season is not suitable for selection of straw strength, plant height, earliness and resistance to some diseases and shattering. Dry seasons occur frequently. Hence breeder has to wait for normal season for effective selection for above characters in wheat. Similar examples can be cited from many other crops. Harland used mass pedigree method in cotton.

Q7. Explain briefly achievements of pedigree method.

Ans. Pedigree method has been extensively used for developing improved varieties in several self pollinated crops like wheat, rice, pulses, barley, cotton and various vegetable crops. Some examples are given below :

- ✰ *Rice*: Varieties Krishna, Ratna, Sabarmati, Padma, Jaya, Bala, Kaveri, *etc.*
- ✰ *Wheat*: K65, K68, NP52, NP120, NP125, NP700, *etc.*
- ✰ *Cotton*: J34, J205, LH372, H655C, SH131, MCU8, MCU9, Sujay, Jayadhar, Raichur 51, Suvin, *etc.*
- ✰ *Pigeon pea*: T21, Prabhat
- ✰ *Green gram*: T2, T44, T51. Sheela, *etc.*
- ✰ *Chickpea:* T1, T2, T3, T5, Radhey, *etc.*

Q8. What are the merits and demerits Pedigree method?

Merits

(i) Pedigree method provides information about the mode of inheritance of various qualitative characters which is not possible by other breeding methods.

(ii) There are chances of recovering transgressive segregants by pedigree method.

(iii) This method takes 14-15 years to release a new variety whereas bulk method takes much longer time (15-16 years).

(iv) The breeding value of selected plants is ascertained by progeny test. Thus pedigree selection is based on genotypic value rather than phenotypic value.

Demerits

(i) The selected material becomes so large that handling of the same becomes very difficult

(ii) Records have to be maintained for all the selected plants and progenies which take lot of valuable time of a breeder.

(iii) Since large number of progeny are rejected in this method, there are chances of elimination of some valuable material.

Q9. Describe briefly main features of bulk breeding method.

Ans. Bulk breeding refers to a selection procedure in which the segregating population of self pollinated species is grown in bulk plot (From *F1* to *F5*) with or without selection, a part of the bulk seed is used to grow the

next generation and individual plant selection is practised in *F6* or later generations. This method is also termed as mass or population method. This method was developed by Nilsson Ehle in 1908. The main features of bulk breeding method are given below:

(i) **Application.** This method is used for the genetic improvement of self pollinated crop plants. It is used when both the parents are adapted or have good agronomic characters and the character for which improvement is to be made is governed by polygenes.

(ii) **Handling of material.** The material is handled by bulk method from *F2* to *F5* and by individual plant selection as in pedigree method from *F6* onwards. The bulk period varies from 6-10 years in short term bulks and

(iii) **Selection.** In this method both natural as well as human selection operate. Natural selection operates during bulk period and human selection operates in the later generations when individual plant selection is practised.

(iv) **Adaptation.** This method leads to significant evolutionary changes in the gene frequencies in a population. Hence, it is also referred to as evolutionary method of crop improvement. Natural selection in favours those genotypes which have better survival capacity and eliminates genotypes with poor survival capacity. Thus, proportion of the best fit genotypes will increase and of poorly adapted genotypes decrease. The best surviving genotypes sometimes may not be agronomically good. Hence, artificial selection should be practised during bulk period to avoid drastic changes in genes; and genotypes through natural selection. Varieties developed by bulk method are more stable against environmental changes than those evolved by pediigree breeding method, because the period of bulking improves the adaptationi of population.

(v) **Genetic constitution.** The end product of bulk breeding method is a homozygous arid homogeneous population, because it is the progeny of a single homozygote.

Q10. Describe briefly breeding procedure of bulk breeding method.

Ans. The bulk breeding method consists of four important steps, *viz.* (i) bulking period, (ii) progeny selection and isolation of superior progenies, (iii) multilocation trials of superior progenies, and (iv) release of best progeny as a variety. These steps are briefly discussed below :

(i) **Bulk period.** The *F1* plants are grown and their *F1* seeds are harvested in bulk. The *F2* plants are raised from a sample of *F2* seeds and *F3* seeds are harvested in bulk. This process is repeated until the desired level of homozygosity is achieved. In general, bulk period is allowed upto *F5* generation (Figure 7.2). The material is subjected to biotic and abiotic stresses during bulking period to eliminate undesirable genotypes.

(ii) **Progeny selection.** In *F6*, the material is space planted and individual plant selection is practised. The progeny of each selected plant is grown separately in *F7* and superior progenies are selected and isolated in *F7* and *F8*.

(iii) **Multilocation testing.** The selected progenies constitute strains. In eighth year, preliminary yield trial is conducted. From 10th to 14th year multilocation testing is carried out and the best performing strain is identified on the basis of 3-4 years performance in the multilocation trials. The best strain is released and multiplied for seed distribution in the 15th year. Thus bulk method takes 15-16 years for release of new variety are isolated based on preliminary replicated trial. The superior progenies are then tested in multilocation trials and the best progeny is identified for release.

The bulk material should be tested under such environment which is expected to favor desirable genotypes. For example, screening for disease resistance should be carried out in the disease prone area. Moreover, the screening should be done under natural field conditions rather than in greenhouses.

Q11. Explain briefly the merits and demerits of bulk breeding method.

Merits

(i) Bulk breeding is a simple, convenient and less expensive method of crop improvement.

(ii) In this method natural selection operates which results in elimination of undesirable genotypes from the bulk population and increases the frequency of desirable plants.

(iii) The chances of obtaining transgressive segregants are more in this method than pedigree method, because the material is grown in large plots in this method.

(iv) This method is useful to study the competitive ability of various genotypes in the population from 20-30 years in long term bulks. In crop breeding, a 5-6 year period of bulking is usually adopted.

(v) This method leads to improvement in adaptation of genotypes due to bulk period. Thus varieties developed by this method are more adaptable than those evolved by pedigree method.

(vi) There is no need to maintain pedigree records in this method.

Demerits

(i) This method does not provide information about the mode of inheritance of various oligogenic characters which is obtained in pedigree method.

(ii) The long term bulking requires 20-30 generations and short term bulking requires 6-10 generations. Thus this method takes more time than pedigree method in release of new variety. For this reason, this method is not preferred by the plant breeders.

(iii) It is difficult to assess the variability in the population and genotypic frequencies in this method, because they change in each generation of bulking.

(iv) Sometimes, natural selection may favor undesirable than desirable genotypes,

Q12. Explain briefly achievements bulk breeding method.

Ans. This method has been used only in few crops due to time lag. It has been used in the improvement of barley in USA and more than 50 varieties have been developed from composite crosses by this method. In India, only one variety "Narendra Rai" has been developed in Brown Mustard by this method. Thus bulk method has limited application in practical plant breeding.

Q13. Explain briefly main features of single seed descent method.

Ans. A breeding procedure used with segregating populations of self pollinated species in which plants are advanced by single seeds from one generation to the next is referred to as single seed descent method. This method was suggested by Goulden (1939) for advancing segregating generation of self pollinated crops. Later on this method was applied by Grafius (1965) in oats, Brim (1966)

(i) Application. This method is used for the genetic improvement of self pollinated crop plants. This method has been used in soybean, wheat, barley, oats, rice, chickpea, green gram and some other crops.

(ii) Handling of material. In this method, only one seed is selected randomly from each plant in *F2* and subsequent generations. The selected seed is bulked and is used to grow the next generation. This process is generally continued up to *F5* generation.

(iii) Selection. In this method natural selection operates during bulk period and human selection operates in the later generations when individual plant selection is practised.

(iv) Adaptation. Varieties developed by single seed descend method are more stable against environmental changes than those evolved by pedigree breeding method, because the period of bulking improves the adaptation of population.

(v) Genetic constitution. The end product of bulk breeding method is a homozygous arid homogeneous population, because it is the progeny of a single homozygote.

Q14. Explain briefly the breeding procedure of single seed descent method.

Ans. This is a modified form of bulk breeding method. In this method, only one seed is selected randomly from each plant in *F2* and subsequent generations. The selected seed is bulked and is used to grow the next generation. This process is generally continued up to *F5* generation. By this time desired level of homozygosity is achieved. In *F6*, large number of single plants (400 to 500) are selected and their progenies are grown separately. In *F7* and *F8*, selection is practiced

Q15. What are the merits and demerits of single seed descent method?

Merits

(i) This is a simple, convenient, less expensive and time saving method. There is no need of keeping much record in this method.

(ii) Large number of crosses can be evaluated by this method, because less space and labor is required in each generation.

(iii) This method is able to retain considerable variability in a breeding population.

(iv) The material can be rapidly advanced by growing the same in greenhouse or off season nursery.

Demerits

(i) This method does not provide opportunity to practise selection for superior plants till F5 generation. Thus many superior plants may be lost.

(ii) The frequency of getting desirable genotypes in the advanced generation is reduced in this method.

(iii) The identity of superior F_2 plants cannot be maintained in this method, their identity is lost.

(iv) This method is more useful when several generations can be grown per year.

(v) This method is applicable to self pollinated crops only.

Q16. Explain briefly main features of backcross method.

Ans. Backcross refers to crossing of F_1 with either of its parents. When the F_1 is crossed with homozygous recessive parent, it is known as test cross. A system of breeding in which repeated backcrosses are made to transfer a specific character to a well adapted variety for which the variety is deficient is referred to as backcross breeding. The main features of backcross method of breeding are briefly presented below:

(i) **Application.** The backcross method is generally used to improve specific character of a well adapted variety for which it is deficient such as resistance to a specific disease. This method is more commonly used for transfer of monogenic or oligogenic characters than polygenic characters. In other words, it is more successful when the character has high heritability. Oligogenic characters have high heritability than polygenic traits. This method is commonly used in self and cross pollinated species. In vegetatively propagated crops like sugarcane and potato this method is rarely used and that too with some modifications.

(ii) **Parental material.** Backcross method involves two types of parents, *viz.*, recipient parent and donor parent. The parent which receives a desirable character is known as recipient parent. The recipient parent is repeatedly used in the backcross method, hence it is also called as recurrent parent The recipient parent is generally a well adapted high yielding variety of an area which is deficient in one or few characters. The parent which donates the desirable character is known as donor parent. Since donor parent is used only once in the crossing, it is also known as nonrecurrent parent. The donor parent is generally poor in agronomic characters. Thus backcross method is used when one of the parents is unadapted type.

(iii) **Genetic constitution.** Backcross method retains the genotype of original variety except for the character which is improved by backcrossing. In other words, the new variety resembles the parent variety in all the characters except for the character under transfer.

(iv) **Number of back crosses.** Generally 5 to 6 backcrosses are sufficient to retain the genotype of original variety with new character.

(v) **Basic requirements.** The basic requirements to start a backcross programme are (i) recurrent parent, (ii) donor parent, and (iii) high heritability of the character under transfer.

Q17. Explain briefly the genetic basis of backcrossing.

Ans. Backcross increases the frequency of desirable individuals in a population. For example, from a cross involving single locus (*AAxaa*), we will get only 1/4 desirable individuals (*AA*) in F_2 through selfing (1AA : 2*Aa*: 1*aa*). In case of backcrossing (AA x *Aa*), we get 1/2 desirable individuals in the BC *F*1 (1AA : 1*Aa*). The same thing is expected for each gene pair. The population gradually becomes identical to the recurrent parent. The population is not divided into 2″ homozygous genotypes as happens in case of selfing. However, in backcrossing homozygosity is attained at the same rate as with selfing which is given below:

Proportion of homozygous individuals = [(2m - 1)/2m]n

where m = number of backcrossing or selfing and

n = number of gene pairs.

Moreover, the chances of breaking linkage between desirable and undesirable genes are more with backcrossing than with selfing. Suppose, gene *A* is desirable and it is linked with undesirable gene *b*. The desirable gene A has to be transferred from a donor to a well adapted variety. The cross between adapted and donor parents will produce *AaBb* hybrid. The genes A and 'a' have the tendency to inherit together to make it difficult to obtain *AB* combination. Since gene *B* is reintroduced with each backcross, there will be several chances for the crossover to take place. Thus the probability of elimination of *b* gene is as given below:

Probability of eliminating of b gene = 1 - (1 - p)m+1,

where p = recombination fraction and m = number of backcrosses

Q18. Explain briefly the breeding procedure of backcross method.

Ans. Some characters are governed by dominant gene and others by recessive gene. The breeding procedure of backcross method depends on whether the character under transfer is controlled by dominant or recessive gene. The breeding procedure for both the situations is briefly presented below:

(i) Transfer of dominant gene: Suppose wilt resistance in cotton is controlled by a dominant gene (*RR*). The donor parent is a strain (*B*) from the germplasm. The resistance has to be transferred to an adapted variety (*A*) which is susceptible to wilt. The adapted variety (*A*) will be used as recurrent parent and strain (*B*) as donor parent. The F_1 will be wilt resistant but heterozygous (*Rr*). Backcrossing of F_1 (*Rr*) with susceptible variety (*rr*) will produce resistant and susceptible plants in equal number in BC_1F_1 (1*Rr* : 1rr). The resistant cotton plants (*Rr*) can be identified by growing the material in wilt sick plot. The resistant plants (*Rr*) are then backcrossed to the adapted variety. Generally, 6-8 backcrosses are sufficient to obtain identical plants to adapted variety except Selection is practiced for the added genes for wilt resistance. The wilt resistant plants are heterozygous (*Rr*). They are selfed for one generation to obtain homozygous (*RR*) resistant plants. All the resistant true breeding plants are bulked and new variety is released. The variety developed in this way is identical to the adapted variety (*A*) except for wilt resistance (Table 7-2) between progenies and superior progenies.

TABLE 7-2. Transfer of Wilt Wesistance in Cotton Governed by Single Dominant Gene [RR] from a Donor Line [A] to a well Adapted but susceptible variety [A]

Year/ Season	*Backcross Generation*	*Crosses between Female Male*		*Resulting Genotypes and Generation Genotypes Generation*		*Recovery of Original Variety (per cent)*
1	-	rr [A]	RR [B]	Rr	F1	50.000
2	BC1	rr [A]	Rr	Rr : rr	BC1 F1	75.000
3	BC2	rr [A]	Rr	Rr :rr	BC2 F1	87.500
4	BC3	rr [A]	Rr	Rr :rr	BC3 F1	93.750
5	BC4	rr [A]	Rr	Rr : rr	BC4 F1	96.875
6	BC5	rr [A]	Rr	Rr :rr	BC5 F1	98.438
7	BC6	rr [A]	Rr	Rr :rr	BC6 F1	99.218
8	BC7	rr [A]	Rr	Rr : rr	BC7 F1	99.805
9	BC8	rr [A]	Rr	Rr :rr	BC8 F1	99.902
10	BC9	rr [A]	Rr	Rr :rr	BC9 F1	99.951
11	BC10	rr [A]	Rr	Rr :rr	BC10 F1	Self to get RR plants

(ii) **Transfer of recessive gene:** Suppose wilt resistance in cotton is governed by a recessive gene (*rr*). In such case, the progeny of each backcross will segregate into two genotypes (*RR* and *Rr*) which cannot be identified. Therefore, it is necessary to self the population after each backcross to obtain resistant homozygous recessive plants (*rr*). The resistant plants are identified by growing the F_2 material in wilt sick plot. The resistant plants are backcrossed with adapted variety. Here each backcross is followed by one selfing, whereas with dominant gene continuous backcrosses are made.

(iii) **Transfer of quantitative traits:** Backcross method is generally used for transfer of monogenic or oligogenic characters. It can also be used for transfer, of polygenic traits. However, transfer of polygenic characters is somewhat difficult due to low heritability of such characters and more influence of environment in the expression of polygenic characters. For successful transfer of polygenic character, the non recurrent parent with extreme phenotype for the polygenic character under transfer should be chosen. For example, if we want to improve protein percentage from 20 to 25 per cent , we should select non recurrent parent with 30 per cent protein. This will make identification of the character easy. Moreover, after each backcross one or two generations of selfing are required, to get the desirable segregants. Furthermore, large populations have to be raised to achieve the desired combination. In other words, the observations should be based on large samples.

Sometimes, several characters have to be transferred into an adapted cultivar through backcrossing. This can be achieved in two ways (1) transfer of genes in separate backcross program and then combining them into single genotype, and (2) simultaneous transfer of genes into single genotype in one backcross program. For simultaneous transfer of multiple characters, backcross seeds have to be produced in more quantity to be sure to get a genotype with all desirable genes.

Q19. Give a brief comparison of pedigree and backcross methods of breeding.

Ans. A comparison of pedigree and backcross methods of breeding is presented in the following Table.

TABLE 7-3. Comparison of Pedigree and Backcross Methods of Breeding

Sl.No.	*Particulars*	*Pedigree Method*	*Backcross Method*
1	Application in	Self pollinated crops	Both self and cross pollinated crops
2	Crossing is done	Once	Repeatedly
3	Selection applied	Human selection	Human selection
4	Size of F2 population to be evaluated	Larger than backcross method	Smaller than pedigree method
5	Pedigree records	Maintained	Not maintained

Sl.No.	Particulars	Pedigree Method	Backcross Method
6	Effectiveness	Equal for both oligogenic and polygenic traits	More effective with oligogenic traits
7	Time taken to release new variety	14-15 years	7-8 years
8	Adaptation of variety developed	Narrow	Wide
9	New variety is	Different from both parents	Differs from recurrent parent for one character only
10	Breeding procedure for dominant and recessive traits	Is same	Differs
11	Use of method	Widely used	Widely used

Q20. What are merits and demerits of backcross method of breeding?

Merits

(i) Backcross method retains all desirable characters of a popular adapted variety and replaces undesirable allele at a particular locus.

(ii) This is a useful method for transfer of oligogenic character like disease resistance. It is also useful in the incorporation of genes for quality such as protein content.

(iii) This method is extensively used in the development of varieties with multiple disease resistance. Multiline varieties carrying resistant genes for different races of a pathogen are also developed by backcross method. This is used for development of isogenic lines and multiline variety, a mixture of several isogenic lines.

(iv) The male sterility and fertility restorer genes are transferred to various agronomic bases by this method.

(v) This is the only breeding method which is used for interspecific gene transfer.

(vi) The variety developed by this method does not require multilocational testing, because it is identical to parent variety except for the character under transfer.

Demerits

(i) This method is used to rectify the defect of an adapted variety. The new variety differs from the old one only in respect of defect which has been rectified.

(iii) It involves lot of crossing work. The backcrosses have to be made for 6-8 generations. In pedigree and bulk methods hybridization is done only once.

(iii) Sometimes, undesirable character is tightly linked with desirable one, which is also transferred to the new variety.

Q21. Explain briefly main achievements of backcross method of breeding.

Ans. Backcross method has been widely used for the development of disease resistant varieties in both self and cross pollinated species. It has also been used for interspecific gene transfer and development of multiline varieties in self pollinated species. Several varieties resistant to various diseases have been developed by this method in wheat, cotton and several other crops. In cotton varieties V797, Digvijay, Vijalpa and Kalyan which belong to *Gossypium herbaceum* have been developed by backcross method.

Q22. What is multiline breeding? Describe its main features.

Ans. The deliberate seed mixtures of isolines, closely related lines or unrelated lines are referred to as multi-lines, and a variety which is developed for commercial cultivation from any of these mixtures ***Backcross Method*** is known as multiline variety. The terms multilines and blends are used as synonyms. However, some persons prefer to designate mixtures of isoline as multiline and mixtures of lines differing for several characters as blends. Isogenic lines or isolines refer to those lines that are genetically identical except for the allele at one locus. In other words, isogenic lines have only one gene difference. The use of multiline cultivars was first suggested in oats by Jensen in 1952. Borlaug and Gibler in 1953 outlined the method for developing multiline in wheat. Later on several workers used this method in various self-pollinated species.

Main features of multiline varieties are briefly presented below:

1. **Application:** The multiline approach is applicable to self-pollinated sepeices only. Multiline cultivars are commercially used in self/pollinated crops like oat, wheat, soybean, goundnut and many other crops.
2. **Genetic Constitution:** Multiline cultivars are mixtures of several pure lines. The pure lines may be isogenic lines, closely related lines or unrelated lines. Thus, multilines are homozygous but heterogeneous populations or genetically diverse populations. The genotypes which are mixed together to constitute a multiline have phenotypic similarities for several characters like height, maturity, grain color and size *etc.*
3. **Adaptation:** Multilines are more adaptable to environmental variations than purelines by virtue of their genetic diversity. In other words, multilines have more buffering capacity to environmental changes than pure lines. The pure lines are adapted to specific environment but have poor adaptability. Multilines have broad genetic base which provides them greater adaptability.
4. **Disease Control:** The use of multiline cultivars is an effective way to minimize the yield losses due to the attack of multiracial disease. In a multiline cultivar, each component genotype has a resistant gene for a different race of a disease. All races of a disease will never appear at a time and all the gonotypes of a heterogeneous mixture are never attacked at a time. Multilines cause delay in the spread of the pathogen in the field,

because the resistant plants act as barriers in the spread of a disease. In this way multiline cultivars provide better protection from the attack of new race of a disease.

5. **Quality of Produce:** The produce of multiline cultivars is generally less uniform and less attractive than that of a pure line, because it is a mixture of several pure lines.
6. **Yield:** The yield of a multiline would be lesser than that of the most productive cultivar of a pureline under normal conditions. But under adverse conditions, the yield of a multiline would be much higher than that of most productive pure line cultivar. Because highly productive pure line cultivars are more prone to biotic and abiotic stresses due to narrow genetic base.

Q23. What are types of multiline?

Ans. There are three types of multiline: *viz.* (1) mixtures of isolines, (2) mixtures of closely related lines, and (3) mixtures of unrelated or distinctly different genotypes. These are briefly described below:

1. **Mixtures of Isolines:** Isolines are genotypes having one gene difference only. Isolines are developed by backcross method. The resistant genes for different races of a disease are transferred into one popular variety by separate backcross programmes. Six to ten isolines are developed and their seeds are mixed in equal quantity to constitute a multiline. Such varieties have been developed in oat, soybean and wheat in USA.
2. **Mixtures of Closely Related Lines:** Sometimes, multiline cultivars are constituted by mixing the seed of closely related lines. Closely related lines are developed from crosses having one parent in common. This approach of multilines development is being followed now-a-days at important breeding centres like CIMMYT. A mixture of closely related lines of wheat KSML 3 was released from Ludhiana to provide better resistance to rust disease (Gill *et al.*, 1980). The six components of multiline were derived from crosses with popular cultivar Kalyan Sona as the common parent. Different types of crosses were made to develop each of the components including single crosses and limited backcrossing. In this case the component lines are not isolines. They differ in several characters from each other.
3. **Mixtures of Unrelated Lines or Cultivars:** Multilines are also constituted from seed mixtures of distinctly different cultivars. Such multilines are developed when phenotyipc uniformity is not essential. First purelines are developed by pedigree, bulk or single seed descent methods. The performance of each pure line is evaluated and then superior pure lines each with different gene for resistance are mixed together to constitute multiline. This is almost similar to varietal blends.

Q24. Describe in brief the breeding procedure of developing multiline.

Ans. The development of a multiline consists of four important steps : (1) selection of recurrent parent, (2) selection of donor parents, (3) transfer of resistant genes into recurrent parent, and (4) mixing of seed of the isogenic lines. These are briefly discussed below :

1. **Selection of Recurrent Parent:** The recurrent parent should be a high yielding popular variety. The recurrent parent should be the best cultivar of a region.
2. **Selection of Donor Parents:** Parents with resistance to various races of a disease should be chosen as donor parents. The resistance should be thoroughly examined under artificial epiphytotic conditions before use of the donor parents in the crossing programs. The donor parents should be adapted varieties as far as possible. Because in un-adapted parents disease resistance is sometimes linked with several undesirable characters and transfer of resistant genes from such parents to the recurrent parent becomes difficult task. Several donor parents are selected to incorporate different resistant genes against various races.
3. **Transfer of Resistance:** The resistant genes are transferred from donor parents to the recurrent parent through a series of several separate backcross programmes. Generally 4-5 backcrosses are sufficient to retain the genotype of recurrent parent with added resistance in the backcross derivatives. The backcross derivatives are evaluated for disease resistance during backcrossing and also at the end of backcrossing. The desirable lines from each backcross are mixed to form an isoline.
4. **Mixing of Isolines:** The various isolines developed by various backcrosses are mixed together to constitute a multiline cultivar. Generally 6-10 isolines are mixed to constitute a multiline cultivar.

Q25. Explain briefly the merits and demerits of multiline variety.

Merits

1. Multiline are more adaptable to environmental changes than pureline cultivars due to genetic diversity.
2. They provide better protection form the infection of new race of a disease.

Demerits

1. The produce of multiline varieties is less attractive and less uniform due to mixture of several pure lines.
2. Development of multiline cultivars involves several backcrosses and hence is costlier than conventional breeding methods.

Multiline cultivars have been developed for commercial cultivation in oats, wheat, soybean and peanut in USA. In India three multiline varieties, *viz.* KSML 3, MLKS 11 and KML 7404 have been released in wheat from Punjab. The first two varieties involve 8 closely related lines and the third one involves 9 closely related lines.

Chapter 8

Breeding Cross Pollinated Species (Mass, Progeny and Ear to Row Selection)

Q1. What are the breeding methods that are used for genetic improvement of cross pollinated crops?

Ans. Breeding methods that are used for genetic improvement of cross pollinated or allogamous species include: (1) Plant Introduction, (2) Mass and progeny selection, (3) Backcross method, (4) Heterosis breeding, (5) Synthetic breeding, (6) Composite breeding, (7) Polyploidy breeding, (8) Distant hybridization, and (9) Transgenic breeding. Mutation breeding is rarely used in allogamous species. Three breeding approaches, *viz.*, recurrent selection, disruptive mating and bi-parental mating are used for population improvement. Various selection schemes which are commonly used for genetic improvement of cross pollinated species include, mass selection, progeny selection, line breeding, ear to row selection, recurrent selection, synthetic breeding, composite breeding and heterosis breeding.

Q2. What mass selection? Describe its main features.

Ans. Mass selection is one of the oldest methods of crop improvement. In this method, individual plants are selected on the basis of phenotype from a mixed population, their seeds are bulked and used to grow the next generation.

The main features of varieties developed by mass selection are given below:

1. **Application:** Mass selection is applicable to both self and cross pollinated species. However, it is more commonly used in the improvement of cross

pollinated crops than in self pollinated species. This method is rarely used in vegetatively propagated crops.

2. **Genetic constitution.** In self pollinated crops, a mass selected variety is homozygous but heterogeneous, because it is a mixture of several purelines. In cross pollinated crops, such varieties are mixture of homo and heterozygotes and are heterogeneous, because they consist of several homo and heterozygous genotypes.
3. **Adaptation.** Mass selected varieties have wide adaptation and are more stable against environmental changes due to heterogeneity which provides better buffering capacity. In other words, mass selected varieties have broader genetic base than pure lines. They exhibit more or less stable performance. However, adaptability is more in cross pollinated crops than in self pollinated species.
4. **Variation.** They are composed of several pure lines in self pollinated crops and of several homo and heterozygous genotypes in cross pollinated crops. Hence there is heritable variation in the mass selected varieties, besides environmental variation. The heritable variation provides them good buffering capacity.
5. **Selection.** Selection is effective in case of mass selected varieties of self pollinated crops due to presence of heritable varieties. However, further selection in the mass selected varieties of cross pollinated crops may lead to inbreeding depression.
6. **Quality.** A variety developed by mass selection is less uniform in the quality of seed than pure lines due to presence of heritable variation.
7. **Resistance.** Mass selected varieties are less prone to the attack of new diseases due to genetic diversity. In other words, they are more resistant or tolerant to new diseases.
8. Periodic removal of off type plants is essential to maintain the yield of mass selected varieties.

Q3. What are the types of mass selection?

Ans. There are two types of mass selection, *viz.* (1) positive mass selection, and (2) negative mass selection. These are defined below :

1. **Positive mass selection.** When desirable plants are selected from a mixed population and their seeds are mixed together to grow further generation, it is referred to as positive mass selection. This process is continued for several years. Generally, old varieties or land races are used as the base material for mass selection. Selection of desirable plants or positive approach is in common use in mass selection.
2. **Negative mass selection.** When only undesirable off type of plants are removed from the field and rest are allowed to grow further, it is known as negative mass selection. This is generally used for varietal purification in seed production and certification programs. This helps in maintaining

high level of genetic purity in the varieties especially in the self pollinated species.

The success of mass selection mainly depends on three factors, *viz.* (1) variability in the base population, (2) mode of inheritance of character to be improved, and (3) heritability of the character. Mass selection is more successful in old heterogenous variety or land races than in improved varieties.

Q4. What are the modifications of mass selection?

Ans. Mass selection is practiced both in self and cross pollinated crop plants. However, it is more common in cross pollinated species than in self pollinated species. There are two defects of mass selection as given below:

1. **No Control on Pollination:** In mass selection there is no control on the pollination. The selected plants are pollinated both by superior and inferior pollen parents.
2. **Selection is based on Phenotype:** The phenotypic performance is greatly influenced by environmental factors such as soil heterogeneity.

In order to overcome these defects, three modifications of mass selection have been suggested in cross pollinated species. These modifications are; (i) rejection of inferior pollen plants, (ii) use of composite pollen, and (iii) stratification of the field. A brief description of each method is presented below:

(i) **Rejection of Inferior Plants.** In this method inferior pollen parents are removed before flowering and rest are allowed for intermating. This controls pollination by inferior plants.

(ii) **Use of Composite Pollen.** In this method, pollen is collected from the selected plants and bulked. This composite pollen is used to pollinate the selected plants. This also controls pollination by inferior plants.

(iii) **Stratification of Field:** This is known as grid method of mass selection and was suggested by Gardner (1962). In this method, the field in which mass selection is to be practiced is divided into small plots in such a way that each plot should have 40-50 plants. Now superior plants are selected in each small plot. This eliminates the effect of environment [soil heterogeneity] because small plots are more homogeneous than larger plots.

The success is more with oligogenic recessive characters than with polygenic characters. In case of polygenic traits, large number of individual plants should be selected in each generation to tap maximum desirable alleles. Moreover, mass selection is more effective for characters having high heritability. Generally, oligogenic characters have higher heritabilily than polygenic characters.

In cross pollinated species, the progeny of each cycle of mass selection should be grown in isolation to prevent contamination with other varieties (mating types) of a crop. The isolation distance differs from crop to crop depending upon the agency of pollination. Generally, more isolation distance is required for wind pollinated species than insect pollinated ones. In general, isolation distance of 300 metres in case of insect pollinated species and 500-1000 metres for wind pollinated species is sufficient. In self pollinated species distance of 5-10 meters is adequate.

Q5. Explain briefly the procedure of mass selection.

Ans. Mass selection consists of various steps, *viz.* selection of base population, selection of desirable plants from base population and mixing their seeds to raise next generation, evaluation in field trials, and releasing as a new variety. The general procedure of mass selection is given below :

1. **First year.** An unimproved old variety or land race is used as a base population which is grown in a large plot Then individual plants (55-1000) are selected on the basis of phenotypic performance for characters like height, maturity, disease resistance, productivity *etc.* The selected plants are harvested at maturity and their seeds are mixed together to grow next generation. This process is repeated till desirable results are achieved.
2. **Second year.** The crop is grown from the bulk seed of selected plants in a separate field using standard variety as a check for comparison of performance. In other words, the material is evaluated in preliminary yield trial. If mass selection is used for purification of old mixed variety, the same old variety can be used as check for comparison.
3. **Third to sixth year.** The performance of bulk is evaluated for yield and adaptation in main yield trials for 3 to 4 years using standard check for comparison.
4. **Seventh and eight year.** The variety is released and named in seventh year and seed is multiplied. In the eighth year the seed is ready for distribution.

Q6. What are the merits and demerits of mass selection?

Merits

(i) This is a good method for improvement of old varieties and land races. This is also used for the purification of improved varieties.

(ii) Mass selected varieties are more stable in their performance than pure lines. In other words, they have more buffering capacity than pure lines due to heterogeneity.

(iii) Mass selected varieties provide good protection against diseases.

(iv) Mass selection is a simple and quick method of crop improvement. It takes about 8 years for the release of a new variety, whereas pure line selection takes about 10 years in the development of new variety.

(v) This method is applicable to both self and cross pollinated species.

Demerits

(i) The selection is based on the phenotypic performance. The superior phenotype is not always an indication of superior genotype. The real breeding value of single plants can be judged from the performance of their progeny. Progeny test is not carried out in mass selection.

(ii) In cross pollinated species, there is no control on the pollination. The selected plants are pollinated by both superior and inferior pollen parents. This results in rapid deterioration of variety developed by mass selection.

(iii) In cross pollinated crops, large number of plants have to be selected for bulking, because small sample will lead to inbreeding depression.

(iv) The produce of varieties developed by mass selection is less uniform man pure lines. This is because mass selected varieties are mixture of several pure lines in self-pollinated crops and consists of several genotypes in cross pollinated species.

(v) In self- pollinated species, pure line selection is more effective than mass selection, because pure line selection leads to isolation of best line from a mixed or heterogeneous population.

Q7. What are the achievements of mass selection?

Ans. Mass selection has played more significant role in developing new varieties in cross pollinated species than in self pollinated species. In India, mass selection has been useful in the development of improved varieties in cross pollinated crops like maize, pearl millet and mustard, and in often cross pollinated species like cotton and sorghum. In self pollinated species it has been rarely used because pure line selection is more effective in such crops than mass selection.

Q8. What is progeny selection? Describe its main features.

Ans. A selection procedure in which superior plants are selected from a heterogeneous population on the basis of the performance of their progeny is referred to as progeny selection.

The genetic worth of an individual is assessed by progeny test: The test of the genotypic value of an individual based on the performance of its progeny is called progeny test. The progeny test was developed by Louis de Vilmorin, hence it is also known as Vilmorin principle. The progeny test is useful in two ways, *viz.* (i) In understanding whether a plant is true breeding (homozygous) or segregating (heterozygous) for a particular character, and (ii) in the assessment of the breeding value of a plant. If the selected plant is really superior, its progeny will also exhibit superior performance and vice versa. Main features of progeny selection are given below:

1. **Application:** Progeny selection is commonly used in cross pollinated and often cross pollinated species.

2. **Base Material**: In cross pollinated species, three types of materials, *viz.* (i) open pollinated seeds, (ii) self seeds, and (iii) top cross or test cross seeds of selected plants can be used in producing the progenies for testing. Generally, 10-50 seeds of each selected plant are grown for progeny testing.
3. **Basis of Selection**: In this method, selection of superior plants is based on the progeny performance (genotype).Those plants whose progeny performance is superior for the character under consideration are bulked together to produce the next generation
4. **Conduct of Progeny Test:** Progeny test should be conducted in replicated trial to get more reliable results. The multi-locational or multi-seasonal test is considered ideal for progeny testing. Finally, plants whose progeny performance is good are selected and rest are discarded.
5. **Genetic constitution.** A variety developed by progeny selection is heterogeneous, because it consists of several homo and heterozygous genotypes.
6. **Adaptation.** A variety developed by progeny selection has wide adaptation and is more stable against environmental changes due to heterogeneity which provides better buffering capacity.
7. **Variation.** They are composed of several homo and heterozygous genotypes. Hence there is heritable variation in the varieties developed by progeny selection. The heritable variation provides them good buffering capacity.
8. **Selection.** Further selection in a variety developed by progeny selection may lead to inbreeding depression.
9. **Quality.** A variety developed by progeny selection is less uniform in the quality of seed than pure lines due to presence of heritable variation.
10. **Resistance.** A variety developed by progeny selection is less prone to the attack of new diseases due to genetic diversity.

Q9. Give a comparison of mass selection and progeny selection.

Ans. A comparison of mass selection and progeny selection is presented in the following Table.

TABLE 8-1. Comparison of Mass Selection and Progeny Selection

Sl.No.	*Particulars*	*Mass Selection*	*Progeny Selection*
1	Applicable to	Both Self and cross pollinated crops	Cross and often cross pollinated crops
2	Selection is based on	Phenotypic performance	Progeny performance
3	Adaptation of variety	Wide	Wide
4	Control over pollination	No control	No control
5	Genetic Composition	Heterogeneous	Heterogeneous
6	Selection	Effective after 4-5 generations	Effective after 4-5 generations

Sl.No.	Particulars	Mass Selection	Progeny Selection
7	Time taken to release new variety	8-10 years	8-10 years
8	Vulnerability to new race of a disease	Low	Low
9	Exact reconstitution	Not possible	Not possible
10	Maintained by	Open pollination	Open pollination

Q10. What are the merits and demerits of progeny selection?

Merits

1. Selection is based on progeny performance which is a reliable technique.
2. This method is very simple and convenient.

Demerits

1. The main demerit of progeny selection is that there is no control on the pollination. The plants of superior progeny are also pollinated by the plants of inferior progenies.
2. If the progeny of each plant is tested in isolation, it would require lot of area which is not practically possible.

Q11. What is ear to row selection? Describe its main features.

Ans. Ear to row selection may be defined as a method of plant breeding in which selection is carried out on progenies derived from individual ears. The concept of ear to row selection was developed by Hopkins in 1908 working with maize crop in USA. Main features of ear to row selection are briefly presented as follows:

1. **Application:** This breeding method was first used by Hopkins in maize. Now this technique is commonly used for genetic improvement of maize. It is the simplest form of progeny selection. Hence it is applicable for the improvement of both self and cross pollinated crops.
2. **Base Material**: In cross pollinated species various breeding materials such as an open pollinated variety, composite, synthetic variety, a mass selected variety, a single cross hybrid, a three-way cross hybrid or double cross hybrid can be used as base material to start ear to row selection. In self- pollinated species, progeny row selection can be used with released variety.
3. **Adaptation**: The variety developed by ear to row selection has broad genetic base, because it is composed of several homo and heterozygotes. Thus, it has wide adaptation.
4. **Basis of Selection**: Initially, the plants are selected on the basis of phenotype. Final selection is made on the basis of progeny performance of progeny rows or ear rows.

5. **Variation:** There is genetic variation in the variety developed by ear to row selection, because such variety is a mixture of several homozygotes and heterozygotes. Thus, selection is effective in such variety.
6. **Quality of Produce**: Since a variety developed by ear to row method is a mixture of several homozygotes and heterozygotes, the produce of such varieties is not uniform.
7. **Maintenance:** The variety developed by ear to row method is maintained by open pollination. However, periodical removal of off types is required for maintenance of such variety.

Q12. Describe briefly the breeding procedure of Ear to row selection.

Ans. Hopkins in 1908 developed this method, extensively used in maize. In its simplest form it consists of following steps:

1. **Selection of Plants**: About 50- 100 number of plants are selected from heterozygous source on the basis of their phenotype and are allowed to open pollinate. The Ears from individual plants are harvested separately.
2. **Planting Ear to Row Progeny**: A single row of 10-50 plants *i.e.* a progeny row, is grown from each selected plant. The progeny rows are evaluated for desirable character and superior are identified.
3. **Selection of Superior Progenies**: Several phenotypically superior plants are selected from the superior progenies and selected plants are permitted to open pollinate.
4. **Second Cycle**: Small progeny rows are again grown from the selected plants, and the process of selection is repeated.

Q13. Describe briefly the merits and demerits of ear to row selection.

Merits

1. In this method the final selection is based on the progeny test and not phenotype of individual plants, hence it is more efficient than mass selection in the identification of superior genotypes.
2. Inbreeding may be avoided to certain extent by selecting sufficiently large number of progenies.
3. It is simple and easy.

Demerits

1. There is no control on pollination and plants are allowed to open pollinate, thus selection is based on maternal parent only. This reduces the efficiency of selection.
2. Many progeny selection schemes are complicated and involve considerable work.
3. The selection cycle is usually of the two years. Thus, time requirement for selection is twice as that of mass selection.

A number of modifications have been suggested to overcome pollination by inferior plants. Some modifications are presented as follows:

1. The selected progenies may be selfed instead of open pollination. This will avoid pollination by inferior pollen.
2. The selected plants may be crossed to a tester parent. The tester parent may be an open pollinated variety, or an inbred. This will improve adaptation of the material by creating heterozygosity and heterogeneity in the material.
3. The progeny test may be conducted in replicated trial. This will help in improving precision of the results.

Q14. What is line breeding? Explain briefly its main features.

Ans. A system of breeding in which a number of genotypes with superior performance for several characters are composited to form a variety is known as line breeding. Main points related to line breeding are given below:

1. **Application:** Line breeding approach is commonly used in animal breeding and less frequently in plant breeding. In plant breeding it is used mostly in cross pollinated species.
2. **Approach**: line breeding can involve crossing between closely or distantly related lines, but it does not emphasize continuous inbreeding.
3. **Purpose:** The main purpose of line breeding is to transmit a large proportion of desirable genes from generation to generation without causing an increase in the frequency of undesirable traits often associated with inbreeding.
4. **Genetic Constitution**: In cross pollinated species, a variety developed by line breeding approach consists of several homo and heterozygotes and thus has genetic diversity.
5. **Adaptation**: A variety developed by line breeding approach has wider adaptation due to presence of genetic diversity.
6. **Reconstitution**: Reconstitution of a variety developed by line breeding approach is not possible due to change in the gene frequencies.
7. **Maintenance**: A variety developed by line breeding approach is maintained by open pollination which permits pollination by both superior and inferior pollen plants.

Q15. What are the merits and demerits of line breeding?

Ans. Intensive inbreeding (and resulting increased homozygosity) is often directly related to an increase in the expression of many undesirable traits. Line breeding is usually associated with little risk of producing undesirable individuals. Line breeding does not necessarily cause a rapid increase in homozygous gene pairs. Consequently, it will not expose undesirable recessive genes as extensively as inbreeding. For this reason, line breeding is generally a safer program for improvement of cross pollinated species

The main demerit of this approach is that there is no control over the pollination. The pollination is effected by both superior and inferior pollen producing plants.

Chapter 9

Clonal Selection

Q1. What are clonally propagated crop plants?

Ans. Those crops which are propagated asexually or by vegetative means are known as asexually propagated or vegetatively propagated or clonal crops. Some examples of asexually propagated crop plants are given below:

1. Agricultural Crops: Sugarcane, potato, sweet potato, *etc.*
2 Horticultural Crops: Banana, mango, citrus, apple, pears, peaches, loquat, litchi, *etc.*

Q2. What are the reasons of asexual reproduction?

Ans. The main reasons of asexual reproduction are: (i) reduced flowering and seed set, (ii) non flowering in many cases, (iiii) to avoid inbreeding depression in certain crops, and (iv) apomixis in some species.

1. **Non flowering species :** This group includes garlic, ginger, betel and several yams.
2. **Low seed setting species :** Sugarcane, potato, sweet potato *etc.*
3. **Normal flowering and seed setting species :** Citrus, mango, pear, peach, apple, litchi, loquat and many ornamental plants. These are highly heterozygous and vegetative propagation is essential to maintain the heterozygous balance.
4. **Apomictic species :** Seeds develop asexually in such species. Many fruit crops exhibit apomixis.

Q3. Explain main features of clones.

Ans. Progeny of a single plant obtained by asexual reproduction is known as clone. A procedure of selecting superior clones from the mixed population of asexually propagating crop is referred to as clonal selection.

Clones have several important features such as: (i) homogeneous constitution, (ii) heterozygosity, (iii) vigorous growth, (iv) wider adaptation *etc.* These are briefly discussed below:

1. **Homogeneous constitution:** The progeny of a clone is genetically identical. In other words, all the plants of a clone have similar genetic constitution. Thus clones are homogeneous. There is no genetic variation within a clone. The variation is only environmental hence selection is not effective within a clone.
2. **Heterozygosity:** The asexually propagated crops are heterozygous and hence clone is also heterozygous. Progeny of a clone looks similar phenotypically but is heterozygous. If a clone is subjected to inbreeding, it will produce various types of segregants and exhibit high inbreeding depression.
3. **Vigorous growth:** Clones have hybrid vigour which is conserved due to asexual reproduction. Most of the varieties of sugarcane and potato are hybrids. In other words, clonal selection is useful in conserving the heterosis for a long period, because clones are stable and are not prone to segregation.
4. **Wider adaptation:** Generally, clones are more adaptable to environmental variation due to high level of heterozygosity than pure lines. A deliberate mixture of genetically different but phenotypically similar clones give better yield in variable environments than a single clone and also provides better protection from the infestation of diseases.
5. **Source of variation:** There are three main sources of variation in a clone, *viz.*, bud mutations, mechanical mixtures, and occasional sexual reproduction. The frequency of bud mutation is very low. But once bud mutation occurs it will lead to deterioration of a clone by adding new variants in the population. Viral and bacterial diseases also lead to deterioration of a clonal variety.
6. **Segregation in F_1:** When hybridization is done between different clones, segregation occurs in F_1 generation. Each F_1 plant is potentially a new variety, therefore, selection is practiced in F_1.

Q4. Give a brief comparison of a pure line variety and a clonal variety.

Ans. A brief comparison of a pure line variety and a clonal variety is presented in Table 9.1.

Table 9-1. Comparison of a Pureline and a Clonal Variety

Sl.No.	*Particulars*	*A Pure line*	*A Clone*
1	Relevant to	Self pollinated crops	Asexually propagated species
2	Genetic constitution	Homozygous and homogeneous	Heterozygous and homogeneous
3	Adaptation	Narrow	Wide

Sl.No.	Particulars	A Pure line	A Clone
4	Quality of produce	Highly uniform	Highly uniform
5	Component genotypes	Identical	Identical
6	Selection	Not effective	Not effective
7	Time taken to release new variety	8-10 years	8-10 years
8	Adaptation of variety developed	Narrow	Wide
9	Vulnerability to new race of a disease	High	High
10	Exact reconstitution	Not possible	Not possible
11	Maintenance by	Natural self pollination	Asexual propagation

Q5. Define clonal selection and describe its main features.

Ans. Progeny of a single plant obtained by asexual reproduction is known as clone. A procedure of selecting superior clones from the mixed population of asexually propagating crop is referred to as clonal selection. The Crops which are propagated asexually or by vegetative means are known as asexually propagated or vegetatively propagated or clonal crops. The main features of clonal selection are presented as follows:

1. **Relevance:** Clonal selection is relevant to vegetatively propagated or asexually propagated crops sugarcane, potato, sweet potato and some fruit crops.
2. **Base material:** The mixed population of vegetatively propagated crops is used as base material to start clonal selection.
3. **Variation:** In a clonal variety, the variation is entirely due to environmental factors, because all the genotypes of such varieties are identical. In other words, there is no genetic variation within a clone. The variation is only environmental hence selection is not effective within a clone.
4. **Genetic constitution:** A variety developed by clonal selection is heterozygous but homogeneous. All the plants of a clonal variety have exactly the same genetic constitution. The progeny of a clone is genetically identical. In other words, all the plants of a clone have similar genetic constitution. Thus clones are homogeneous.
5. **Adaptation:** A clonal variety is more adaptable to environmental variation due to high level of heterozygosity than pure lines. A deliberate mixture of genetically different but phenotypically similar clones give better yield in variable environments than a single clone and also provides better protection from the infestation of diseases.
6. **Disease resistance:** Clonal varieties are highly vulnerable to new race od a disease due to narrow genetic base.
7. **Reconstitution:** Exact reconstitution of a clonal variety is not possible.

8. **Produce:** The produce of a clonal variety is highly uniform due to homogeneous nature.
9. **Maintenance:** The clonal variety is maintained by asexual propagation.

Q6. Describe briefly then breeding procedure of clonal selection.

Ans. Improvement of asexually propagated crops by selecting superior clones is known as clonal selection. Superior clones can be isolated from three types of material, *viz.* (i) local variety, (ii) introduced variety, and (iii) intercrossed populations. In other words, clonal selection can be practised in above three types of material. The general procedure of clonal selection is outlined below :

1. **First Year**: In the first year, superior plants are selected from a mixed population of vegetatively propagated crop. Superior plants are selected on the basis of yield, maturity, disease resistance, *etc.*
2. **Second Year**: In the second year, the progeny of each selected plant is asexually propagated and grown separately for seed increase and evaluation. Clones superior to check variety are selected and evaluated in replicated preliminary trials in the third year.
3. **Third to Sixth Year**: Several clones are evaluated in multilocation trial for 3-4 years using standard variety as a check. Clones which are superior to check variety in yield, disease resistance and quality are identified for release.
4. **Seventh to Eighth Year**: The best clone is released as a variety and its seed is multiplied for distribution.
5. **Ninth Year**: Distribution of seed to the farmers for commercial cultivation. Thus release of new variety by this method takes 9-10 years.

Besides clonal selection, inter-specific hybridization and mutation breeding are also used for the improvement of asexually propagated crops. These methods have been successfully used in sugarcane and potato. Inter-specific hybridization has been extensively used in the breeding of sugarcane. Many modem cultivars have been derived from crosses of *saccharam officinarum* with *S, spontaneum* or *S. barberi.* These crosses are useful in combining high sugar content of the first species with disease resistance, cold tolerance and vigour of last two species. Backcross method is used for transfer of desirable gene from wild species to the cultivated one in sugarcane and potato.

Q7. What are merits of clonal selection?

1. This method is useful in conserving heterosis for several generations. The variety evolved by this method retains all the characteristics of the parental clone for several years.
2. Clonally propagated varieties are highly uniform like purelines. Moreover, they are highly stable because there is no risk of deterioration due to segregation and recombination.

3. This is an effective method for genetic improvement of asexually propagated crop plants. In other words, it is useful in isolating best genotype from a mixed population of asexually propagated crops.

Q8. What are demerits of clonal selection?

1. Varieties developed by clonal selection are highly prone to new races of a disease.
2. Clonal selection cannot create new variability and, therefore, genetic makeup cannot be improved by this method without hybridization.

Q9. Give a brief account of practical achievements of clonal selection.

Ans. In India, clonal selection has been successfully used for developing new varieties in potato, sugarcane, banana, citrus and grapes. Varieties Kufri Red and Kufri safed in potato; Ko 11, Ko 22 and Neelam in mango; and Bombay green, Pride monthan and High gate in banana have been developed by clonal selection. Several varieties of sugarcane (Co 541, CoS 510, Co 1148, Co 1158, *etc.*) and potato (Kufri, Sinduri, Kufri Kuber, Kufri Kundan, Kufri Chamatkar, *etc.*) have been developed by interspecific hybridization followed by clonal selection.

Chapter 10

Transgenic Breeding

Q1. What is transgenic technology?

Ans. The technique or process of developing transgenic individuals is referred to as transgenic technology. It is also known as gene technology, transformation technology, genetic engineering and recombinant DNA technology. The crop plants which are developed through the application of plant biotechnology or plant genetic engineering are called transgenic plants or genetically modified plants or genetically engineered plants. In broad sense transgenic plants can be called as genetically modified plants. But in strict sense, they are genetically engineered plants. However, the term GM crops or GM plants or GM organisms has been very popular and is widely used the world over. The GM plants have been developed to solve specific problems which could not be solved easily through conventional plant breeding methods.

Q2. What are transgenic plants?

Ans. The genetically engineered plants are called transgenic plants. Main features of transgenic plants are listed as under.

(i) Transgenic plants contain transgene or foreign gene of special significance.

(ii) Transgenic plants are developed through plant biotechnology.

(iii) In the development of transgenic plants, the sexual process is not involved.

(iv) The transgenic plants are recovered at a very low frequency.

(v) Transgenic plants are developed to solve those problems which cannot be solved by conventional plant breeding techniques.

Q3. What are advantages of transgenic technology?

Ans. Transgenic technology has several advantages such as rapid method, free gene transfer, single gene transfer, direct gene transfer and solution to difficult problems. These are briefly discussed as follows:

1. **Rapid Method:** It is an effective and rapid method of crop improvement. It takes 3-4 years for developing new cultivars/hybrids against 10-15 years taken by conventional methods.
2. **Free Gene Transfer:** It permits gene transfer across the species, genera, family even from unrelated organisms.
3. **Single Gene Transfer:** Gene technology permits transfer of one or two genes from donor parent to the recipient parent. In conventional hybridization method, hundreds of genes are transferred to the recipient parent. Many of transferred genes are undesirable. Elimination of such genes requires repeated backcrossing to the recipient parent which takes 4-5 years.
4. **Direct Gene Transfer:** Gene technology permits direct gene transfer into the recipient parent bypassing sexual process. In other words, there is no need of union of male and female gametes in gene technology. The gene of interest can be directly inserted into the cell of recipient parent.
5. **Solution to Difficult Problems**: Gene technology provides solution to those problems that cannot be solved by conventional methods of breeding. The best example is resistance to bollworms in cotton.

Q4. What are applications of transgenic technology?

Ans. Transgenic technology has several practical applications. Important applications of transgenic technology include improvement in (i) biotic resistance, (ii) abiotic resistance, (iii) herbicide resistance, (iv)quality of food products and (v) development of novel traits such as golden rice and male sterility. All these aspects are briefly discussed as follows:

1. **Resistance to Biotic Stresses:** Biotic stress refers to adverse effects on crop growth and yield by biotic factors such as insects, diseases and parasitic weeds. In crop plants, heavy yield losses are caused every year due to insect and disease attack. Moreover, insecticides and pesticides which are used to control insects and diseases are expensive and have adverse effects on other beneficial organisms (parasites and predators). Gene technology has played key role in developing insect resistant cultivars in several crops such as bollworm resistant cultivars in cotton and stem borer resistant cultivars in maize. Moreover, the technology is eco-friendly.

2 **Resistance to Abiotic Stresses**: Abiotic stress refers to adverse effects on crop growth and yield by abiotic factors such as drought, soil salinity, soil acidity, cold, frost, *etc.* The cold resistant genotypes in tobacco and freezing resistant genotypes have been developed through gene technology. Efforts are being made to develop drought and salinity resistant cultivars in many crops.

3. **Herbicide Resistance:** In crop plants, weeds cause heavy yield losses and also adversely affect quality of the produce. The genetic resistance is the cheapest and the best way of solving this problem. Gene technology has been used to develop herbicide resistant cultivars in cotton, maize, wheat, tobacco, potato, tomato, rapeseed, soybean, flax, *etc*. In these crops, cultivars resistant to glyphosate, gluphosinate and some other herbicides have been developed.
4. **Improvement in Quality:** The quality is adjudged in three ways, *viz.*, nutritional quality, market quality [keeping quality] and industrial quality. Gene technology has helped in improving these qualities in different crops. For example, the ripening and softening in tomato has been delayed. It is desirable for safe transport and storage. This has been achieved by manipulating the genes that encode the enzyme responsible for ripening [ethylene forming enzymes and softening [polygalactonase]. Other quality improvement include non-browning potato, starch composition of wheat flour, carotene content in rice and improved oil content in oilseed crops.
5. **Development of Novel traits:** Plant biotechnology plays important role in development of novel traits. Examples are golden rice and transgenic male sterility in rapeseed.
6. **Industrial Products:** Gene technology has great potential for the production of biodegradable plastics, obtaining therapeutic proteins, pharmaceuticals and edible vaccines from transgenic plants. It may also help in producing biodiesel or petroleum products.
7. **Bioremediation:** It is possible to develop plants which can be used for bioremediation of sick soils.

Q5. What are practical achievements of transgenic technology?

Ans. Plant biotechnology has played significant role in (i) improving resistance [to insects, diseases, drought, salinity, frost, cold, *etc.* (ii) food quality and (iii) inducing male sterility in some crops. Some examples of such achievements are cited below [Table 9-2].

1. **Improvement in Resistance:** It includes insect resistance, disease resistance, salinity resistance and herbicide resistance. Such resistance has been achieved in several crops as mentioned below:

   (i) **Insect Resistance**: It includes bollworm resistance in cotton, European borer resistance in maize and Colorado beetle resistance in Potato. Resistance to these insects has been transferred from *Bacillus thuringiensis*.

   (ii) **Disease Resistance**: Resistance to Pierce disease in grapes has been developed at the University of Florida. Resistance to mosaic in papaya, potato and squash has been achieved in USA.

(iii) **Salinity Resistance**: Salinity resistant tomato has been developed at the University of California and at the University of Toronto.

(iv) **Herbicide Resistance**: It has been achieved in cotton, Flax, Soybean, Wheat, Potato, Tomato, Rapeseed, *etc.* The gene for herbicide resistance has been transferred from *Steptomyces* to wheat, potato and tomato; and from microbial gene in cotton, soybean and linseed.

2. **Improvement in Quality**: Quality improvement has been achieved in several crops. Some examples are cited below:

(i) **High Starch Potato**: In potato starch contents have been increased by transferring gene from human intestine bacteria [*Escherichia coli*].

(ii) **High Protein Potato and Alfalfa**: Protein contents have been improved in potato and alfalfa. In potato, the high protein gene has been transferred from Amaranth and serum albumin gene from human. In alfalfa ova-albumin gene has been transferred from chicken.

(iii) **High Methionine Soybean**: In soybean, methionine content has been enhanced by transferring gene from a bacteria [*Bertholletia excelsa*].

(iv) **High Carotene Rice**: In rice, high carotene content gene has been transferred from Daffodils and **Golden rice** has been developed. The golden rice is rich in vitamin A content. Two genes have been transferred from Daffodils and one from bacterial species.

(v) **Anti-cracking Tomato**: In tomato, anti-cracking gene has been transferred from winter flounder fish. This has increased the transportation and keeping quality of tomato.

(vi) **Freezing Resistant Tobacco**: In tobacco, freezing resistant gene has been transferred from winter flounder fish and cold resistant gene from *Arabidopsis thaliana*.

3. **Male Sterility:** The male sterility has been developed in rapeseed through the application of agricultural biotechnology. The *barnase* and *barstar* system has been developed in this crop. The male sterility gene has been transferred from *Bacillus amyloliquefaciens.*

TABLE 10-1. Traits Improved in some Genetically Modified Crop Plants

Sl.No.	*Traits Improved*	*Examples of GM crops*
I	**Resistance**	
1	Insect Resistance	Cotton, Corn, Tobacco, Potato, *etc.*
2	Disease Resistance	Potato, Tobacco, Squash, Grapes, *etc.*
3	Freezing Resistance	Tomato, Tobacco, *etc.*
4	Salinity Resistance	Tomato
5	Herbicide resistance	Cotton, Corn, Soybean, Rapeseed, Alfalfa, Linseed, Sugar beet,
II	**Quality**	
1	Starch content	Potato
2	Protein content	Potato, Alfalfa, *etc.*

Sl.No.	Traits Improved	Examples of GM crops
3	Methionine	Soybean
4	Carotene content	Golden Rice
5	Oleic acid content	Soybean
6	Male sterility	Rapeseed

Q6. What are the traits improved using transgenic technology?

Ans. Several plant traits in GM crops have been improved through the application of plant biotechnology and different gene sources have been used for such improvement. Some examples are given in Table 10-2

TABLE 10-2. Traits Improved and Gene Sources Used in some GM Crops

Sl.No.	Crop	Traits Improved	Gene Used From
1	Soybean	Herbicide resistance Methionine	Microbial gene *Bertholletia excelsa*
2	Maize	Insect resistance [ECB] Herbicide resistance	Bt. Gene *Arabidopsis thaliana*
3	Cotton	Bollworm resistance Herbicide resistance	Bt. Gene Microbial gene
4	Canola	Male sterility Herbicide resistance	*Bacillus amylolifaciens* *Arabidopsis thaliana*
5	Papaya	Ring spot virus	Wild species
6	Squash	Mosaic virus	Coat protein gene
7	Potato	Protein content Mosaic resistance Herbicide resistance	Serum albumin gene of human and amaranth Potato virus x and y Streptomyces
8	Tomato	Insect resistance Freezing resistance Mosaic resistance	Bt. Gene Winter Flounder Fish Tomato mosaic virus
9	Tobacco	Insect resistance Herbicide resistance Cold resistance Freezing resistance	Cowpea TI gene Acaligene Eutrophus Arabidopsis thaliana Winter Flounder Fish
10	Alfalfa	Protein quality	Chicken ova-albumin gene

Q7. What are the probable risks of plant biotechnology?

Ans. There are some probable risks of plant biotechnology such as adverse effects on human and animal health, development of resistance in some weeds against herbicides, ethical issues and development of resistance in insects and diseases against transgenic plants. These are explained as follows:

1. **Adverse Effects:** Consumption of transgenic food and vegetables may have adverse effects on human and animal health such as toxicity and allergenicity. Six transgenic crops, *viz.* soybean, maize, cotton, potato and sugar beet are grown in Australia for human consumption.. However, no adverse effects have been reported so far. The effects of GM crops on health and environment are uncertain. The cultivation of such crops can have unintended adverse effects on both animal health and the environment.
2. **Resistance to Herbicides:** There is fear that continuous application of herbicides may lead to development of resistance in weeds against such herbicides. Development of such resistance has been reported in Johnson grass, Amaranth palmeri and many other weeds against glyphosate herbicide in USA.
3. **Ethical Issue:** The transfer of gene from animal such as from fish to potato and tobacco; ova-albumin gene of chicken to alfalfa and serum albumin gene of human into potato and plants will have ethical problems.
4. **Antibiotic Resistance:** Wide spread cultivation of transgenic plants may lead to development of resistance in insect pests and pathogens against transgenic plants.

Q8. What are the major biotech or GM crops?

Ans. Several GM crops have been released for commercial cultivation in different countries. In Canada, 61 and in USA more than 40 crops have been approved for cultivation. However all have not been adopted by the farmers. The top 10 GM crops grown in different countries include Soybean, maize, Cotton, Rapeseed, Papaya, Squash, Potato, Tomato, Sugar beet and Alfalfa.

Q9. Who are the major producers of biotech or GM crops?

Ans. Biotech or GM crops are grown in more than 30 countries. The top 10 GM crop growing countries include USA, Brazil, Argentina, Canada, India, China, Paraguay, South Africa, Pakistan and Uruguay. The GM crops grown in these countries are presented in Table 14-4. The area of Biotech crops in top 10 countries for three years is presented in Table 4. The maximum area is covered by USA followed by Brazil, Argentina, Canada and India.

Out of above 10 countries, about 70 per cent of GM crop production comes from the first five countries and 90 per cent GM crop area is covered by USA, Argentina, Brazil and Canada.

TABLE 10-3: Area under Biotech Crops in top 6 countries of the world

Sl.No.	Country	2015	2016	2017	2018	GM Crops Grown
1	USA	70.9	72.9	75.0	75.0	Corn, Soybean, Cotton, Canola, Sugar beet, Alfalfa, Papaya and Squash. [8]
2	Brazil	44.2	49.1	50.2	51.3	Soybean, Maize and Cotton [3]
3	Argentina	24.5	23.8	23.6	23.9	Soybean Maize and Cotton [3]
4	Canada	11.0	11.6	13.1	12.7	Canola, Corn and Soybean [3]
5	India	11.6	10.8	11.4	11.6	Cotton [1]
6	China	3.7	2.8	2.8	2.9	Cotton, Papaya, Poplar, Tomato and Sweet Pepper [5]
7	Others	13.8	14.1	13.7	14.3	
	Total	**179.7**	**185.1**	**189.8**	**191.7**	

Other producers of biotech crops are Paraguay (3.4 m ha), South Africa (2.9 m ha), Pakistan (2.8 m ha) and Uruguay (1.4 m ha). Soybean, Maize and Cotton are grown in Paraguay and South Africa, cotton in Pakistan and Soybean and Maize in Uruguay.

Q10. How much Area is under of Major GM crops in top 10 countries?

Ans. The area under GM crops gradually increased from 1996 onwards. In 1996, it was 2.8 m ha which increased to 170.3 m ha in 2012 [Table 6]. Presently, only 10 per cent of global cropped area is under GM crops. Some facts about area under biotech crops are as follows:

(i) Country wise, about 90 per cent area of biotech crops is covered by five countries, *viz.*, USA (41 per cent), Brazil (21 per cent), Argentina (14 per cent), Canada (7 per cent) and India (6 per cent).

(ii) Crop wise, more than 90 per cent area is covered by four crops, *viz.*, soybean (52 per cent), corn (30 per cent), cotton (11 per cent) and Canola (5 per cent). The percent area of these four in main producing countries is presented in Table 10-4.

TABLE 10-4. Area (per cent of total) of Four Main Biotech Crops in Major Producing Countries

Sl.No.	Crop	Area in major producing countries
1	Soybean	USA (98), Canada (70), Argentina (99), Brazil (71), Paraguay (100), South Africa (85), Uruguay ((100), Bolivia (89), Romania (70)
2	Corn	USA (85), Canada (84), Argentina (85), South Africa (63), Brazil (36), Spain (20)
3	Cotton	USA (94), India (93), Australia (95), South Africa (98), Argentina (95),China (80), Pakistan (90), Mexico (56) Mexico (20)
4	Canola	Canada (95), USA (82), Australia (04)

(iii) Character wise, maximum area of biotech crops is covered by herbicide resistance (75 per cent), insect resistance (19 per cent) and remaining by disease resistance quality improvement, *etc.*

(iv) Seven main biotech crops covering maximum area in USA are Corn (88 per cent), Soybean (93 per cent), Cotton (94 per cent), Papaya (75 per cent), Canola (90 per cent) and Sugar beet (54 per cent).

TABLE 10-5. Year-wise Global Area (m ha) under GM Crops from 1996 to 2018

Year	*Area*	*Countries*	*Year*	*Area*	*Countries*
1996	2.8	6	2008	125,0	26
1997	12.0	6	2009	133.0	27
1998	27.8	9	2010	150.0	28
1999	39.9	12	2011	159.0	28
2000	44.2	15	2012	170.0	29
2001	52.6	15	2013	175.2	28
2002	58.7	16	2014	181.5	28
2003	67.7	18	2015	179.7	28
2004	81.0	18	2016	185.1	28
2005	90.0	21	2017	189.8	28
2006	102.0	24	2018	191.7	26
2007	114.0	25			

Q11. What are the top 10 global biotech seed producing companies?

Ans. The seed production of GM crops is carried by global and national private seed companies. The top 10 global seed companies include Monsanto (USA), DuPont (USA), Syngenta (Switzerland), Groupe Limagrain (France), Land cy Lakes (USA), KWS AG (Germany), Bayer Crop Science (Germany), Sakata (Japan), DLF Trifolium ((Denmark) and Takaii (Japan). In India, there are more than 30 seed companies engaged in GM crop production. The top 10 companies include Mahyco, Monsanto, Nuziveedu, Rashi seeds, Syngenta India, Bayer Crop Science, Ankur Seeds, Ajeet Seeds, Nath Seeds and Bioseeds.

Q12. What are the Public Sector Organizations in India involved in developing GM crops?

Ans. In India several public sector organizations are involved in development of GM crops. These are as given below:

(i) National Botanical Research Institute (NBRI), Lucknow,

(ii) National Research Centre on Plant Biotechnology (NRCPB), New Delhi,

(iii) International Centre for Genetic Engineering and Biotechnology (ICGEB), New Delhi,

(iv) Central Institute for Cotton Research (CICR), Nagpur,

(v) National Chemical Laboratory (NCL), Pune,

(vi) Bhabha Atomic Research Centre (BARC), Mumbai, and

(vii) University of Agricultural Sciences (UAS), Dharwad.

Q13. Give a brief account of landmarks related to GM crops.

Ans. The first transgenic plant was developed in tobacco in 1983 which was resistant to Kanmycin, an antibiotics. Later on, several GM crops were developed. Important landmarks in the history of GM crops are presented in Table 10-6.

TABLE 10-6. Landmarks in the History of Biotech/GM Crops

Year	*Transgenic/GM Crop Developed*	*Country*
1983	First transgenic plant in Tobacco	USA
1987	Development of Bt. Cotton	USA
1994	FLVR-SAVR Tomato	USA
1994	Bt. Potato	USA
1995	Herbicide resistant Soybean	USA
1996	Herbicide resistant Rapeseed	Canada
1996	Insect resistant Maize [ECB]	USA
1999	Virus resistant Papaya	Hawaii
2000	Golden Rice	Switzerland
2000	Mosaic Resistant Squash	China
2004	High protein Alfalfa	USA
2004	Herbicide resistant Linseed	Canada
2005	Golden Rice 2	USA Syngenta
2008	Mosaic resistant Papaya	Thailand
2009	Herbicide resistant Sugar beet	Canada
2012	Mosaic free Banana	S. Africa and Australia
2013	Drought tolerant sugarcane	Indonesia
2014	Modified starch potato	USA
2015	Late blight resistant potato	USA
2015	HR Soybean	Canada

Q14. Give a brief account of research in advanced stage related to GM crops.

Ans. In order to solve some important problems, the research work on GM crops is in advance stage on the following aspects:

1. Development of protein packed potatoes.
2. Development of Nicotine free tobacco.
3. Development of vitamin E rich Canola.
4. Development of banana with inactivated viruses which cause cholera, hepatitis B and diarrhoea.

5. Improvement in the storage and keeping quality of fruits and vegetables.
6. Production of oral vaccines and other pharmaceuticals in plants.
7. Production of biodegradable polymers for the plastic industry.
8. Plants that can be used for bioremediation of polluted soils.
9. Forest trees with enhanced growth and improved timber properties.
10. Production of enhanced quality of chemicals for chemical industry.

Q15. What are the future prospects of transgenic technology?

Ans. Remarkable work related to GM crops has been done in the past on improving insect resistance, disease resistance, herbicide resistance, quality improvement and several transgenic cultivars have been released for commercial cultivation in crops like soybean, corn, cotton, canola, *etc.* It is expected that plant biotechnology [GM crops] can revolutionize world agriculture, particularly in developing countries. This technology will substantially improve food security, reduce malnutrition, and increase rural income. In future, the transgenic research on development of GM crops needs to be directed towards the following thrust areas and crops.

1. **Improvement in Resistance:** There is need to develop GM crop cultivars resistant to drought and soil salinity in countries like India.
2. **Nutritional Quality:** There is need to develop GM crop cultivars of fruits and vegetables with improved nutritional, market and keeping qualities.
3. **Crops to be Improved:** It is expected that in future [by 2015], the main focus would be on the development GM cultivars in rice (stem borer), brinjal (fruit borer), potato (mosaic resistance), and wheat (herbicide resistance).

   There is no need of developing herbicide resistant crop cultivars, because there is huge application of herbicides in weed control which pollutes the soil, underground water and has adverse effects on soil micro-organisms [earth worms, N_2 fixing bacteria, blue green algae, *etc.*]

Q16. What are the points which should be given due importance in future related to transgenic technology?

Ans. In future, to achieve agricultural revolution through gene technology, the following points should be given due importance.

1. **Affordable Technology:** The new technology should be affordable by the farmers of developing-world. Moreover, the farmers must understand how to use them.
2. **Research by Public Sector:** There is a need for larger investments in research in the public sector. Partnerships between the public and private sectors can result in more efficient production of GM crops that are useful to the developing world and can expand the accessibility of those crops and their associated technologies to developing-world farmers.

3. **Due Importance to Agriculture:** Policy makers should give due importance to agriculture to fight hunger and malnutrition in near future particularly in under developed and developing countries.
4. **Proper Regulatory System:** Policymakers in the developing world must set regulatory standards that take into consideration the risks as well as the benefits of foods derived from GM crops. Proper regulatory system is essential to curb malpractices associated with seeds of GM crops.
5. **Specific Applications:** Plant biotechnology should be used only to solve those problems that cannot be addressed by conventional breeding techniques or to create novel genotypes for various purposes.

Q17. Give a comparison of transgenic and conventional breeding.

Ans. The genetic improvement of crop plants through the application of agricultural biotechnology [transgenic plants] is called transgenic plant breeding. Transgenic crop breeding differs in many ways from conventional methods of crop breeding. Comparison of transgenic breeding and traditional plant breeding is presented in Table 10-7.

TABLE 10-7. Comparison of Transgenic Breeding and Traditional Breeding Methods

Sl.No.	*Particulars*	*Transgenic Technology*	*Traditional Breeding*
1	Used in	Single celled organisms	Used in multi-cellular organisms
2	Sexual process	Not Involved	Involved
3	Occurrence in nature	Seldom	Common
4	Crossing barriers	Not present	Present
5	Technique used	Genetic engineering	Conventional breeding methods
6	Methods used	Agrobacterium based method, gene gun, micro injection and direct DNA uptake, *etc.*	Pedigree, bulk breeding, single seed descent, back cross breeding, *etc.*
7	Applications	In all crops	In all crops
8	Bio-safety measures	Required	Not required
9	End product	Transgenic plants	Non transgenic plants
10	Bioethical measures	Required	Not required
11	Evolution of super weeds	Possible	Not possible
12	Organic farming of end product	Not permitted	Permitted
13	Accuracy of method	Very high	Moderate to high
14	Time required for release of new variety	4-5 years	10-12 years
15	Technical skill required	Very high	Moderate
16	Laboratory required	Sophisticated	Ordinary
17	Initial cost of infra structure	Very high	Low
18	Direct single gene transfer	Possible	Not possible
19	Transgene	Involved	Not involved

Chapter 11

SMART Breeding

Q1. What is SMART breeding?

Ans. SMART breeding (SMART=**S**election with **M**arkers and **A**dvanced **R**eproductive **T**echnologies). Smart breeding refers to a process whereby a marker is used for indirect selection of a genetic trait of interest. This is a similar process to traditional breeding, although it makes use of marker assisted selection [MAS]. MAS is a technique that does not replace traditional breeding, but can help to make it more efficient. It does not include the transfer of isolated gene sequences such as genetic engineering, but offers tools for targeted selection of the existing plant material for further breeding. MAS has already proven to be a valuable tool for plant breeders:

(i) It requires less investment, raises fewer safety concerns, respects species barriers, and is accepted by the public. MAS has high potential to meet challenges such as a changing climate, disease resistance or higher nutritional qualities.

(ii) Marker-assisted selection (MAS) is a modern plant breeding technique that can offer benefits to farmers developing climate or diseases resistant varieties, without the need for genetic engineering.

Q2. What are main features of SMART breeding?

Ans. Since smart breeding involves a far more precise knowledge, it is also known as precision breeding. Main features of SMART breeding are listed as follows:

(i) Principle: Smart Breeding is a method based on the results of genomic research. Genomic research can supplant traditional time- and cost-intensive laboratory or field tests and can speed up the selection process.

(ii) Use of Molecular Markers: In smart breeding, when selecting the plants, breeders do not rely on outward features (phenotype). Instead, plant breeders use molecular markers [molecular diagnostic tools - the so-called markers] to select parents for crossing or individual plants with desirable traits.

(iii) Alternative to Transgenic Breeding: In contrast to Genetic Engineering, Smart Breeding does not introduce any new genes into the genome. The resulting organisms are not transgenic. Smart breeding has been advanced as an alternative to transgenic plants as a way to produce plants that are resistant to various environmental problems. Smart Breeding which involves Marker-Assisted selection, is a non-invasive biotechnology alternative to genetic engineering of plant varieties.

(iv) No Effect of Environment: Smart Breeding enables the detection of desired performance features irrespective of environmental influences.

(v) Rapid Method: The analysis of the trait in question is a quick and easily automated process allowing a high throughput. Moreover, the existence of a trait can be analyzed in earlier generations and in earlier development states (germ bud). Traditional methods can only decide upon resistance by carrying out infection trials in breeding nurseries or laboratories and by analysing yield and quality of the harvested crop which is very much time consuming.

(vi) High Accuracy: Smart breeding basically works in a similar way to traditional breeding. However, unlike traditional methods, in smart breeding the gene or gene variant responsible for a specific trait can be accurately identified using molecular biological procedures (DNA sequencing, PCR). It is then possible to test the offspring of a cross for the presence of the crossed gene, even before the actual trait is expressed by a changed external appearance. Only those plants which contain the desired gene are selected and grown further. Thus results of smart breeding are much more predictable.

(vii) Purpose: The purpose of Smart breeding is to introduce into crop plants genes from *e.g.* wild populations which confer characteristics of interest to breeders (disease resistance, fruit color or sugar content).

This technique was successfully used by Nachum Kedar in Israel in tomato breeding to produce a fruit that would ripen on the vine and remain firm in transit.

Q3. What is marker assisted selection? Give its main features.

Ans. Marker assisted selection (MAS) is a process whereby a marker (morphological, biochemical or one based on DNA/RNA) is used for indirect selection of a specific trait. Main points related to marker assisted selection are listed below:

(i) Marker assisted selection is also known as marker aided selection (MAS).

(ii) It is a process in which a marker (morphological, biochemical or one based on DNA/RNA variation) is used for indirect selection of a trait of interest.

(iii) A trait of interest may include productivity, disease resistance, abiotic stress tolerance, quality, *etc.*

(iv) This process is used both in plant and animal breeding.

(v) Marker assisted selection (MAS) is indirect selection process where a trait of interest is selected not based on the trait itself but on a marker linked to it.

(vi) MAS can be useful for traits that are difficult to measure, exhibit low heritability, and/or are expressed late in development.

(vii) MAS can be used as an aid to traditional phenotypic-pedigree-based selection systems.

Q4. Under what situation marker assisted selection is used?

Ans. There are several situations where the use of molecular markers in the selection of a genetic trait would be rewarding. Such situations include (i) late expression of the trait, (ii) trait controlled by recessive gene, (iii) resistance to biotic stresses, and (iv) presence of epistasis. These are briefly discussed as follows:

(i) Late expression of a Trait: MAS will be useful when the trait of interest is expressed late in plant development, like flower and fruit features.

(ii) Control by Recessive Gene: MAS will be useful when the character of interest is is governed by recessive gene. It can identify both dominant and recessive genes.

(iii) Resistance to Biotic Stress: MAS will be useful in screening the material for resistance to diseases and insects especially in those areas field inoculation with the pathogen is not allowed for safety reasons. Moreover, it eliminates undesirable environmental effects.

(iv) Epistasis: MAS will be useful when the phenotype is affected by two or more unlinked genes (epistatis). For example, it is selection for multiple genes which provide resistance against diseases or insect pests for gene pyramiding.

The cost of genotyping (an example of a molecular marker assay) is reducing while the cost of phenotyping is increasing particularly in developed countries thus increasing the attractiveness of MAS as the development of the technology continues.

Q5. What are the steps involved in marker assisted selection?

Ans. The marker assisted selection consists of two important steps, *viz.* (i) mapping of gene, and (ii) use of this information for marker assisted selection. These are discussed as follows:

(i) Gene Mapping: The first step is to map the gene or quantitative trait locus (QTL) of interest. This is done by using different techniques.

Each gene mapping technique has some merits and demerits. Detailed discussion of all such techniques is beyond the scope of this discussion. Five types of populations, *viz.* (i) recombinant inbred lilen(RILs), (ii) near isogenic lines (NILs), (iii) back cross, (iv) double haploids (DH), and (v) F_2 segregations are commonly used for gene mapping or gene tagging. Linkage between the phenotype and markers which have already been mapped is tested in these populations in order to determine the position of the QTL. Such techniques are based on linkage and are therefore referred to as "linkage mapping". This consists of following steps.

1. Selection of parents with divergent alleles or contrasting traits.
2. Development of mapping populations.
3. Isolation of DNA.
4. Scoring of DNA markers such as RFLPs or AFLPs *etc.*

(ii) Marker Assisted Selection: The second important step is to use above information for marker assisted selection. Generally, the markers to be used should be close to gene of interest (<5 recombination unit or cM) in order to ensure that only minor fraction of the selected individuals will be recombinants. Generally, two markers are used in order to reduce the chances of an error due to homologous recombination. For example, if two flanking markers are used at same time with an interval between them of approximately 20cM, there is higher probability (99 per cent) for recovery of the target gene. This consists of following steps.

1. Indirect selection using molecular markers, and
2. Correlation of DNA markers with morphological markers.

Various type of DNA markers are used in marker assisted selection. The commonly used markers include RFLPs, RAPDs, AFLP, SNPs, SCARs, microsatellites *etc.*

Q6. What are applications of markers?

Ans. In most major crops, various linked markers exist to screen key traits. Numerous markers have been mapped to different chromosomes in several crops including rice, wheat, maize, soybean and several others. These markers have been used in diversity analysis, parentage detection, DNA fingerprinting, and prediction of hybrid performance. Molecular markers are useful in indirect selection processes, enabling manual selection of individuals for further propagation. In addition to these, applications of MAS in backcrossing and gene pyramiding are discussed below:

(i) Marker Assisted Backcross: A minimum of five or six-backcross generations are required to transfer a gene of interest from a donor (may not be adapted) to a recipient (recurrent – adapted cultivar). The recovery of the recurrent genotype can be accelerated with the use of molecular markers. If the F1 is heterozygous for the marker locus, individuals with the recurrent parent allele(s) at the marker locus in first or subsequent backcross generations will also carry a chromosome tagged by the marker.

(ii) **Marker Assisted Gene Pyramiding:** Gene pyramiding has been proposed and applied to enhance resistance to disease and insects by selecting for two or more than two genes at a time. For example in rice such pyramids have been developed against bacterial blight and blast. The advantage of use of markers in this case allows select for QTL-allele-linked markers that have same phenotypic effect.

Q7. What are the advantages of marker assisted selection?

Ans. Marker-assisted selection is the most widely used application of DNA markers. Once traits have been mapped and a closely linked marker has been found, it is possible to screen large numbers of samples for rapid identification of progeny that carry desirable characteristics. MAS has several advantages such as (i) high speed, (ii) consistency of results, (iii) bio-safety, (iv) high efficiency, and (v) QTL mapping. These are briefly discussed below:

(i) **Speed:** It is a rapid method of selection of desirable plants. DNA can be extracted from tissue from the first leaves or the cotyledons of a plant. Trait information can be discovered with markers prior to pollination allowing more informed crosses to be made.

(ii) **Consistency:** The results of MAS are consistent, because use of markers eliminates the impact of environmental variation that often complicates phenotypic evaluation.

(iii) **Biosafety:** MAS is a safe method of screening. When we use MAS, there is no need of introducing into breeding populations. Particularly for livestock breeding this delivers a very important level of bio-safety.

(iv) **Efficiency:** MAS has very high efficiency. Screening of progeny in early stages allows a breeder to reject undesirable genotypes from the program more quickly. Most breeding programs that use markers still evaluate the same number of plants in the field however the level of genetic quality is vastly increased because of the early-stage screening that has been carried.

(v) **Complex traits:** DNA markers permit mapping of polygenic traits or quantitative trait loci [QTL] which is not possible through conventional plant breeding techniques.

Q8. What are the disadvantages of marker assisted selection?

Ans. Marker assisted selection has some disadvantages such as (i) expensive technique, (ii) requires technical skill, (iii) a laborious work, and (iv) may sometimes lead to health hazards. These are briefly discussed below:

(i) **Expensive Technique**: MAS is an expensive technique, because this technique requires very costly equipments, glassware and chemicals; sophisticated laboratory.

(ii) **Technical skill:** This technique requires well trained man power for handling of equipments, isolation of DNA molecules and study of DNA markers.

(iii) **Laborious Work**: The detection of various DNA markers [RFLP, AFLP, RAPD, SSR, SNP *etc.*] is a laborious and time consuming work. For such markers huge plant breeding population has to be screened to get meaningful results.

(iv) **Health Hazards**: Some of the DNA marker techniques involve use of radioactive isotopes in labeling of DNA, which may lead to serious health hazards. Now non radioactive labeling agents are also available.

Section III

Seed Production and Crop Breeding

Chapter 12

Varietal Seed Production

Q1. Explain significance of improved seed.

Ans. Production of improved seed is an important area of plant breeding. The seed of a released and popular variety produced by scientific method is referred to as improved seed or quality seed. Improved seed plays an important role in maximizing the production and productivity of field crops. Improved seed leads to (1) better germination, (2) vigorous seedling growth, (3) higher crop stand, (4) better quality of produce, and (5) ultimately in higher crop yield. In plant breeding two types of seed production programs are carried out, *viz.* (i) varietal seed production and (ii) hybrid seed production.

Q2. What are the classes of improved seed? Explain in detail about nucleus seed.

Ans. There are five classes of improved seed, *viz.* (1) nucleus seed, (2) breeder seed, (3) foundation seed, (4) registered seed, and (5) certified seed.

Nucleus seed is the initial seed of an improved variety which is always limited in quantity. It is produced by the originating plant breeder. Main features of nucleus seed are given below :

(i) **Production:** It is produced at the experimental farms of the concerned research institute or agriculture university, under the supervision of original plant breeder or sponsored plant breeder.

(ii) **Purity:** It is genetically and physically cent per cent pure.

(iii) **Certification:** For nucleus seed certification is not required.

(iv) **Uses:** Nucleus seed is used for the production of breeder seed. It is not meant for general distribution.

Q3. Explain briefly about breeder seed.

Ans. Breeder seed is the progeny of nucleus seed or breeder seed. It is produced under the strict supervision of original or sponsoring plant breeder at the research farm of the concerned Crop Research Institute or Agricultural University.

(i) **Production:** Breeder seed is produced in isolation from other varieties. The isolation distance differs from species to species.

(ii) **Purity:** Breeder seed is genetically and physically cent per cent pure. The genetic purity is maintained by proper rogueing.

(iii) **Certification:** Certification is not required for breeder seed. The seed plot is inspected by a monitoring team which consists of original or sponsored plant breeder and one representative each from National Seeds Corporation and State Seed Certification Agency.

(iv) **Uses:** Breeder seed is used for the production of foundation seed. It is not meant for general distribution.

Q4. Describe briefly about foundation seed.

Ans. Foundation seed is the progeny of breeder seed. Its main features are given below:

(i) **Production:** It is produced by the National Seeds Corporation under the strict supervision of research scientists and experts from NSC. Production of foundation seed is taken up at the seed multiplication farms of government, research farms of ICAR Institutes and Agricultural Universities and also on cultivators' fields. Proper isolation distance is adopted for the production of foundation seed which varies from crop to crop.

(ii) **Purity:** Foundation seed is genetically cent per cent pure. However, physical purity of 98 per cent is permissible. The remaining 2 per cent includes inert matter.

(iii) **Certification:** In case of foundation seed, certification is required which is undertaken by the State Seed Certification Agency.

(iv) **Uses:** Foundation seed is used for the production of certified seed. It is not meant for general distribution.

Q5. Give a brief account of registered seed.

Ans. Registered seed is the progeny of either foundation seed or registered seed. In India, registered seed is generally omitted and certified seed is produced directly from the foundation seed. The main features of registered seed are given below :

(i) **Production:** It is produced at the farms of progressive cultivators according to technical advice and supervision of National Seeds Corporation.

(ii) **Purity:** It has 100 per cent genetic purity. However, the physical purity of 98 per cent is permissible.

(iii) Certification: This class of seed also requires certification which is undertaken by the State Seed Certification Agency.

(iv) Uses: Registered seed is used to produce certified seed or registered seed. It is also not meant for general distribution.

Q6. Give a brief account of certified seed.

Ans. Certified seed is the progeny of either foundation or registered or certified seed. Its main features are given below:

(i) Production: It is produced on the fields of progressive farmers under the strict supervision of State Seed Certification Agency. This work is also taken up by the NSC, if required. Proper isolation distance is adopted which varies from species to species.

(ii) Purity: It has genetic purity of 100 per cent and physical purity of 98 per cent. The other crops seeds, and weed seed should not be more than prescribed standards which vary from species to species.

(iii) Certification: This class of seed requires certification which is undertaken by the State Certification Agency. For certification, the seed must meet rigid requirements of purity and germination.

(iv) Uses: Certified seed is available for general distribution to the farmers for commercial crop production.

Q7. Give a brief comparison of breeder and certified seeds.

Ans. The comparison of breeder and certified seeds is presented in Table 12.1.

TABLE 12-1. Differences between Breeder Seed and Certified Seed

Sl.No.	*Particulars*	*Breeder Seed*	*Certified Seed*
1	Progeny of	Nucleus or breeder seed	Foundation or certifies seed
2	Produced by	Original/sponsored breeder	State Seed Corporations
3	Produced at	Research or Govt. farms	Farmers' fields
4	Genetic purity	100 per cent	99.9 per cent
5	Physical purity	100 per cent	98 per cent
6	Certification	Not required	By SSCA
7	Used for	Production of foundation seed	Commercial crop production
8	Color of tag	Yellow	Blue
9	Stage II used	Sometimes	Often

Q8. What are the major steps in quality seed production?

Ans. The production of certified seed differs from crop to crop to some extent. There are some common steps which are involved in the certified seed production of various field crops. The common steps include, (1) package of practices, (2) isolation distance, (3) plant protection measures, and (4) rogueing. These are briefly discussed below:

1. **Package of Practices:** Standard agronomic practices, *viz.* sowing time, spacing, fertilizer dose, hoeing and weeding are available for each crop. These cultural practices have to be strictly followed to raise a good crop. It is essential to harvest good yield and better quality of produce. Hence, recommended agronomic practices should be adopted for production of quality seeds. Proper crop rotation has to be adopted to avoid contamination from previous year's crop variety. The seed production should be taken up under irrigated conditions to ensure high yield and good quality.
2. **Isolation Distance:** Isolation refers to the separation of the field of a variety from that of another variety of the same crop to avoid contamination. Proper isolation distance should be maintained to avoid contamination through natural crossing. The isolation distance differs from crop to crop for the production of foundation and certified seeds (Table 39.2). The isolation distance is low in self- pollinated species, moderate in often cross pollinated species, and high in cross pollinated species. In some self-pollinated species like wheat, barley, oats *etc.*, isolation distance of only 3 m. is required, whereas in cross pollinated species like cabbage, cauliflower, radish, sugar beet *etc.* isolation distance of 1600 m. is required for the production of genetically pure seed.
3. **Plant Protection Measures:** It is essential to protect the seed crop from the attack of various insects and diseases, because attack of both insects and diseases leads to significant

TABLE 12-2(a). Prescribed Standards for Production of Foundation and Certified Seeds of some *Rabi* Field Crops

Sl.No.	*Field Crop*	*Genetic Purity (Minimum)*		*Germination (Minimum)*	
		Foundation Seed	*Certified Seed*	*Foundation Seed*	*Certified Seed*
A	Self-pollinated Species				
1	Wheat	98	98	85	85
2	Barley	98	98	85	85
3	Lentil	98	98	75	75
4	Chickpea	98	98	85	85
5	Field Pea	98	98	75	75
6	Linseed	98	98	80	80
B	Cross Pollinated Species				
7	Mustard/Rapeseed	97	97	85	85
8	Sunflower	98	98	70	70
9	Safflower	98	98	80	80

Note: In case of maize, *Sorghum* and pearl millet isolation distance has been given for open pollinated varieties. For hybrids the isolation distance is still higher. reduction in yield as well as quality of seed. Hence, recommended plant protection measures should be adopted to raise a healthy crop.

TABLE 12-2(b). Prescribed Standards for Production of Foundation and Certified Seeds of some *Kharif* Field Crops

Sl.No.	*Field Crop*	*Genetic Purity (Minimum)*		*Germination (Minimum)*	
		Foundation Seed	*Certified Seed*	*Foundation Seed*	*Certified Seed*
A	Self-pollinated Species				
1	Groundnut	96	96	70	70
2	Black gram	98	98	75	75
3	Green gram	98	98	70	70
4	Soybean	97	97	70	70
5	Cowpea	98	98	75	75
B	Cross-pollinated Species				
6	Maize	98	98	90	90
7	Pearl millet	98	98	75	75
8	Sunflower	98	98	70	70
9	Castor	98	98	70	70
C	Often Cross Pollinated Species				
10	Cotton	98	98	65	65
11	Sorghum	98	98	80	80
12	Pigeon pea	98	98	75	75
13	Sesame	97	97	80	80

4. **Rogueing:** The process of removal of off type (phenotypically different) plants from the field of an improved variety is known as roguing. The main objective of roguing is to avoid contamination through mechanical mixture and due to outcrossing. In self-pollinated crops, generally roguing is done at three different stages, *viz.* before flowering, after flowering and before harvesting. The characteristic features of the variety are taken into account during the process of roguing. Any plant deviating from the features of the variety under multiplication is removed. In the production of hybrid seed in cross pollinated crops, rogueing should be completed before flowering. In self- pollinated species, rogueing can continue even after flowering. The diseased plants should also be removed to prevent the spread of disease. In hybrid seed production, rogueing is done in both seed parent and pollinator parent.

TABLE 12-3: Minimum Field Standards of Some Important Crops

Sl.No.	*Crop*	*Seed Class*	*Isolation*	*Field Inspections*	*Off Types*	*OC Plants*
1	Paddy	FS	3	2	0.05	—
		CS	"	2	0.20	—

Sl.No.	Crop	Seed Class	Isolation	Field Inspections	Off Types	OC Plants
2	Wheat	FS	3	2	0.05	0.01
		CS	150*	2	.20	0.05
3	Sorghum (OP Varieties)	FS	200, 400*	3	0.05** 0.10***	—
4	Pulses	FS	10	2	0.10	—
5	Oilseeds Groundnut	FS	3	2	0.10	—
6	Soybean	FS	3	2	0.10	—
		CS	3	2	0.50	—
7	Cotton	CS	30,5*	2	0.20	—

FS = Foundation Seed, CS = Certified Seed, OC = Other Crop Plants.

Field inspections are made by the inspectors of State Seed Certification Agency to examine the suitability of crop for certification. The number of field inspection varies from 2 to 4 depending upon the crop species. (Table 12-3). When only two inspections are made, one is made during flowering and another before harvesting. In case of three field inspections, one is made before flowering, second at the time of flowering and third before harvesting. In case of four inspections, one is made before flowering, one at the time of flowering and two between flowering and harvesting.

The main objective of field inspections is to examine (1) isolation distance, (2) off types, (3) objectionable weeds, (4) disease and insect incidence, and (5) general crop condition. The off types should not be more than the maximum standard prescribed for. The maximum off type plants in the seed plot of foundation and certified seed plot are permitted from 0.05–1 per cent , depending upon the crop species (for details refer Agarwal, 1980 and Nema, 1986).

Q9. What are the seed producing organizations?

Ans. In India, there are two types of seed producing organizations, *viz.* (1) public sector, and (2) private sector. The public sector organizations include National Seeds Corporation (NSC), State Seed Corporations (SSCs) and State Seed Certification Agencies (SSCAs). Private sector includes various seed companies. A brief account of these organizations is presented below:

1. **National Seeds Corporation:** The NSC was established in March, 1963 to take up the work of quality seed production and promote seed industry in the country. The NSC started functioning in July, 1963. The headquarters of NSC is in New Delhi in the Pusa Campus. The responsibilities of NSC were more clearly defined with the inception of National Seeds Project (NSP) in 1976. The main functions of NSC are listed as follows :

(i) **Production of breeder seed:** The requirement of breeder seed of various crops is assessed and arrangements are made for its production through crop research institutes of ICAR and State Agricultural Universities.

(ii) **Production of foundation seed:** The production and distribution of the foundation seed of the varieties of national importance is the responsibility of the NSC.

(iii) **Production of certified seed:** The production and marketing of the certified seeds of vegetables, oilseeds, pulses and fodder crops is taken up.

(iv) **Marketing.** Interstate marketing of the seeds of national varieties lies with the NSC.

(v) **Consultancy Services:** The NSC provides consultancy services related to designing, procurement, and installation of seed processing plants, equipments, *etc.*

(vi) **Imparts training:** The NSC organizes short-term training courses to the personnel engaged in seed production work.

(vii) **Coordination:** NSC provides coordination in the production of certified seed by SSCs.

(viii) Takes up seed certification work in those states where SSCA has not been established.

(ix) Also takes up import and export work of seed.

2. **State Seed Corporations (SSC):** The National Seed Project was launched in 1976 to establish State Seed Corporations. State Seed Corporations have been established in 14 states. The main functions of the State Seed Corporations are production and distribution of certified seed in the respective state. In Uttar Pradesh, Tarai Development Corporation (TDC) was established in 1969 to produce certified seed which proved very much successful. Looking to the success of TDC in Uttar Pradesh, it was decided to establish SSCs in other states also on the same pattern to faster the development of seed industry in the country. The NSP provides finance for the establishment of new SSCs and strengthening of the existing ones. The NSP also provides funds for the following purposes :

   (i) Strengthening seed production, processing and marketing facilities of NSC.

   (ii) Strengthening the facilities of the NSC for producing and processing seeds of vegetables, fodder crops and other specialized crops not taken up by the SSCs.

   (iii) Strengthening the facilities for nucleus, breeder and foundation seed production; processing, and storage at selected Central Research Institute of ICAR and at SAUs.

3. **State Seed Certification Agencies:** State seed certification agencies are located in 20 different states including those states where state seed

corporations have been established. The main function of the SSCAs is certification of seeds in the respective state. The inspectors of SSCAs make required number of field inspections and conduct purity and germination tests for certification purpose. SSCAs also organize short term training courses for seed growers, keeps check on the source of seed used for raising seed crop *etc.*

Chapter 13

Hybrid Seed Production

Q1. Explain main points related to hybrid seed production.

Ans. The progeny of a cross between genetically dissimilar parents is called hybrid. Hybrid variety refers to those F_1 populations which are used for commercial cultivation. Main points related to hybrid seed production are as follows:

(i) **Application:** hybrid seed production is carried out only for those hybrids which are released either by Central Variety Release Committee or State Variety Release Committee and notified by the Government of India.

(ii) **Production:** Hybrid seed is produced by the National Seeds Corporation and State Seed Corporations. These organizations produce hybrid seed of only those hybrids which have been developed by Government Research Institutes or Agricultural Universities. Private seed companies also produce hybrid seeds of of their own hybrids and also of public sector hybrids.

(iii) **Seed Category**: Hybrid seed belongs to the certified category and, therefore, certification of hybrid seed is essential.

(iv) **Method of Production:** Hybrid seed is produced in two ways, *viz.* by (i) hand emasculation and pollination, and (ii) by using male sterility system. The first method is used for those crops where male sterility system Is not available.

(v) **Seed Used:** Either breeder seed or foundation seed of parental lines should be used for production of hybrid seed. Impurity of parental lines leads to rapid deterioration of hybrid.

Based on cultivation season, hybrids are classified in two groups, *viz.* *kharif* season hybrids and *rabi* season hybrids. *Kharif* season hybrids are developed in maize, Sorghum, pearl millet, pigeon pea, rice and castor.

Rabi season hybrids are developed in rapeseed and mustard, safflower and sunflower.

Q2. What are the steps involved in hybrid seed production?

Ans. Hybrid seed production program differs from crop to crop. However, there are some steps which are common in hybrid seed production of different crops. These steps include: (i) planting ratio, (ii) isolation distance, (iii) rogueing, and (iv) field inspections. These are briefly discussed as follows:

(i) Planting Ratio: The planting ratio of female and male parents differs from crop to crop. Moreover, in some crops female and male parents are planted in adjacent rows, while in other crops they are planted in separate blocks. The recommended planting ratio of female and male parents for some crops is presented in Table 13-1.

TABLE 13-1. Planting Ratio of Female and Male Parents for Hybrid Seed Production in different Field Crops

Sl.No.	*Name of Crop*	*Female : Male*	*Isolation Distance m*	*Planted in*
A	Kharif Hybrids			
1	Maize	6 : 2	200	Adjacent rows
2	Sorghum	4 : 2	100	Adjacent rows
3	Pearl Millet	4 : 2	200	Adjacent rows
4	Pigeon Pea	6 : 1	200	Adjacent rows and adjacent blocks
5	Rice	8 : 2	3	Adjacent rows of A and R lines
6	Cotton	3 : 1	30	Adjacent rows
7	Castor	6 : 2	100	Adjacent rows
B	Rabi Hybrids			
8	Rapeseed and Mustard	4 : 2 or 2 :1	200	Adjacent rows of A and R lines
9	Safflower	6 : 2	200	Adjacent rows and adjacent blocks
10	Sunflower	6 : 2	200	Adjacent rows

(ii) Isolation Distance: Prescribed isolation distance is maintained all around the hybrid seed production plot which varies from crop to crop. However, when hybrid seed production is carried out by hand emasculation and pollination, there is no need of isolation distance such as in cotton.

(iii) Rogueing: Rogueing is essential in female and male parents for pure hybrid seed production. The rogueing should be done at least, one before flowering and second after flowering.

(iv) Field Inspection: The hybrid seed production plots are inspected by the inspectors of State Seed Certification Agency. Number of inspections differ from crop to crop. Field inspections are carried out to examine (a) isolation distance, (b) off types, (c) weeds, (d) insect and disease incidence, and € general crop conditions. All these should be as per prescribe standards.

Q3. What are the factors affecting hybrid seed production?

Ans. Successful production of hybrid seed depend on several factors such as (i) purity of parental lines, (ii) stability of male sterile line, (iii) restoration capacity of R line, (iv) resistance level of parental lines, (v) Yield level of parental lines, (vi) nicking of female and male lines, and (vii) presence of marker in male sterile line. These are briefly discussed as follows:

(i) Purity of Parental Lines: The genetic purity of parental lines is important in hybrid seed production. It leads to stable performance of hybrid over regions and seasons.

(ii) Stability of MS line: Male sterile line should be photo and thermo insensitive. Instability of MS line hamper the hybrid seed production program.

(iii) Restoration Capacity of R Line: The restorer line should be able to completely restore fertility in in F_1. Poor restoration capacity hampers hybrid seed production program.

(iv) Resistance Level of Parental Lines: Parental line should possess high level of resistance to major insects and disease prevailing in the region where hybrid is commercially cultivated. For example in cotton, in north zone leaf curl virus is a major disease. Hence the parents of hybrid should be resistant to leaf curl virus. Several other such examples can be cited.

(v) Yield Level of Parental Lines: The yield level of hybrid depends on the yield level of its parents. Hence, yield level of at least one parent of the hybrid should be high. Moreover, parental line should have broad genetic base which will provide wider adaptability to the hybrid.

(vi) Nicking of Parental Lines: The nicking of flowering period of female and male lines is essential for successful hybrid seed production. If both parents do not flower at one time, the sowing of late flowering parent should be advanced in such a way that their flowering synchronies.

(vii) Presence of marker in MS Line: When hybrid seed production is carried out using genetic male sterile line, 50 per cent of the male fertile plants have to be removed from the female parent. Identification of male fertile plants is possible only after flower initiation. Presence of marker helps in early detection of male fertile plants during seedling stage.

Q4. Describe briefly the main points related to hybrid seed production in Maize.

Ans. The procedure of hybrid seed production differs from crop to crop. Hybrid seed production procedure for some important field crops such as maize, Sorghum, pearl millet, pigeon pea, rice, cotton, castor, rapeseed and mustard, safflower and sunflower is briefly presented as follows:

Maize is a cross pollinated crop in which cross pollination under natural conditions is more than 90 per cent. Cross pollination occurs by wind hence hybrid seed production is easy. In maize, double cross hybrids are common. Hence certified seed of two single crosses is used for production of double cross hybrid. Main points are as follows:

(i) Seed of parents Used: Foundation or certified seed

(ii) Planting Ratio of Female and Male Parent: 6 : 2.

(iii) Isolation Distance: 200 meters

(iv) Rogueing: It should be carefully done both in female and male parents

(v) Agronomic Practices: Recommended agronomic practices have to be followed.

(vi) Plant Protection Measures: Recommended plant protection measures are adopted.

(vii) Field Inspections: Four field inspections are made by seed inspectors of State Seed Certification Agency to check isolation distance, planting ratio, off types, disease and insect incidence and general crop conditions. Detasseling is done when female parent is male fertile.

Q5. Explain main points related to hybrid seed production in Sorghum.

Ans. Sorghum is basically self- pollinated crop in which outcrossing varies from 5-30 per cent. Hence hybrid seed production through natural cross pollination is possible. Main points are as follows:

(i) Seed of parents Used: Foundation or certified seed

(ii) Planting Ratio of Female and Male Parent: 4 : 2.

(iii) Isolation Distance: 100 meters

(iv) Rogueing: Carefully done both in female and male parents

(v) Agronomic Practices: Recommended agronomic practices have to be followed.

(vi) Plant Protection Measures: Recommended plant protection measures are adopted.

(vii) Field Inspections: Four field inspections are made by seed inspectors of State Seed Certification Agency to check isolation distance, planting ratio, off types, disease and insect incidence and general crop conditions.

Q6. Explain main points related to hybrid seed production in pearl millet.

Ans. Pearl millet is a cross pollinated crop in which cross pollination under natural conditions is more than 80 per cent. Cross pollination occurs by wind hence hybrid seed production is easy. In Sorghum, single cross hybrids are made using cytoplasmic genic male sterility. Hence emasculation is not required. Main points are as follows:

(i) Seed of parents Used: Breeder or Foundation seed.

(ii) Planting Ratio of Female and Male Parent: 4 : 2.

(iii) Isolation Distance: 200 metres

(iv) Rogueing: Carefully done both in female and male parents to ensure high genetic purity of hybrid seed.

(v) Agronomic Practices: Recommended agronomic practices have to be followed.

(vi) Plant Protection Measures: Recommended plant protection measures are adopted.

(vii) Field Inspections: Four field inspections are made by seed inspectors of State Seed Certification Agency to check isolation distance, planting ratio, off types, disease and insect incidence and general crop conditions.

Q7. Describe in brief main points related to hybrid seed production in Pigeon pea.

Ans. Pigeon pea is a often cross pollinated in which outcrossing ranges widely (0-70 per cent) under different environmental conditions. The average outcrossing ranges from 12 to 15 per cent. Cross pollination occurs by insects mainly by honey bees and bumble bees. The pollen is heavy and stick hence wind pollination is not possible. Hybrid seed is produced using genetic male sterility. Here 50 per cent male fertile plants in female parent are removed. Main points are as follows:

(i) Seed of parents Used: Foundation or certified seed.

(ii) Planting Ratio of Female and Male Parent: 6 : 1.

(iii) Isolation Distance: 100 meters.

(iv) Rogueing: Carefully done both in female and male parents.

(v) Agronomic Practices: Recommended agronomic practices have to be followed.

(vi) Plant Protection Measures: Recommended plant protection measures are adopted.

(v) Field Inspections: Four field inspections are made by seed inspectors of State Seed Certification Agency to check isolation distance, planting ratio, off types, disease and insect incidence and general crop conditions.

Q8. Describe in brief main points related to hybrid seed production in Rice.

Ans. Rice is a self-pollinated. Cross pollination up to 0.5 per cent is reported. In rice, hybrids are developed through the use of cytoplasmic genic male sterility. Other important points are as follows:

(i) Seed of parents Used: Foundation or certified seed.

(ii) Planting Ratio of Female and Male Parent: 8 : 2. Moreover, R line is planted all around the seed production plot to ensure better seed setting and avoid contamination from other sources.

(iii) Isolation Distance: 3 meters.

(iv) Rogueing: Carefully done both in female and male parents.

(v) Agronomic Practices: Recommended agronomic practices have to be followed.

(vi) Plant Protection Measures: Recommended plant protection measures are adopted.

(v) Field Inspections: Four field inspections are made by seed inspectors of State Seed Certification Agency to check isolation distance, planting ratio, off types, disease and insect incidence and general crop conditions.

Q9. Describe in brief main points related to hybrid seed production in Cotton.

Ans. Cotton is basically a self- pollinated crop in which average outcrossing under Indian conditions is 6 per cent which is not sufficient for hybrid seed production. Hence hybrid seed production is carried out by hand emasculation and pollination. Even if male sterility is available, pollination has to be done by hand Main points are as follows:

(i) Seed of parents Used: Foundation or certified seed.

(ii) Planting Ratio of Female and Male Parent: 4 : 1 or 3 : 1.

(iii) Isolation Distance: 5 meters.

(iv) Rogueing: Carefully done both in female and male parents.

(v) Agronomic Practices: Recommended agronomic practices have to be followed.

(vi) Plant Protection Measures: Recommended plant protection measures are adopted.

(vii) Field Inspections: Four field inspections are made by seed inspectors of State Seed Certification Agency to check isolation distance, planting ratio, off types, disease and insect incidence and general crop conditions.

Q10. Explain briefly main points related to hybrid seed production in Castor.

Ans. Castor is a cross pollinated crop and cross-pollination occurs mainly by wind. Hybrid seed production is carried out by natural cross pollination. In castor hybrid seed is produced by hand emasculation and pollination method, because male sterile line is not available so far.

In castor two types of plants, *viz.* pistillate (female) and monoecious (having both female and male flowers) are found. In the monoecious plants, the raceme bears pistillate flowers in the upper part and staminate in the lower part. Hybrid seed production is carried out using stable pistillate line as female and monoecious line as male parent. Other main points are as follows:

(i) Seed of parents Used: Foundation or certified seed.

(ii) Planting Ratio of Female and Male Parent: 6 : 2.

(iii) Isolation Distance: 100 meters.

(iv) Rogueing: Carefully done both in female and male parents to ensure high genetic purity of hybrid seed.

(v) Agronomic Practices: Recommended agronomic practices have to be followed.

(vi) Plant Protection Measures: Recommended plant protection measures are adopted.

(vii) Field Inspections: Four field inspections are made by seed inspectors of State Seed Certification Agency to check isolation distance, planting ratio, off types, disease and insect incidence and general crop conditions.

Q11. Describe briefly main points related to hybrid seed production in rapeseed and mustard.

Ans. In this group, yellow sarson and mustard are self-pollinated and brown sarson and Toria are cross-pollinated. The outcrossing has been reported from 7.6 to 18 per cent in Indian mustard and up to 16 per cent in *Brassica napus.* Cross pollination occurs through insects and wind. Honey bees are the main pollinating insects. Hybrids are developed using cytoplasmic genic male sterility. Other main points are as follows:

(i) Seed of parents Used: Foundation or certified seed.

(ii) Planting Ratio of A and R Lines: 3 : 1 in *Brassica juncea* and 2: 1 in *Brassica napus.*

(iii) Isolation Distance: 200 meters.

(iv) Rogueing: Carefully done both in female and male parents.

(v) Agronomic Practices: Recommended agronomic practices have to be followed.

(vi) Plant Protection Measures: Recommended plant protection measures are adopted.

(vii Field Inspections: Four field inspections are made by seed inspectors of State Seed Certification Agency to check isolation distance, planting ratio, off types, disease and insect incidence and general crop conditions.

Q12. Describe briefly main points related to hybrid seed production in Safflower.

Ans. Safflower is a cross pollinated crop and cross pollination occurs mainly by wind. Hybrid seed production is carried through the use of genetic male sterility. In GMS, 50 per cent plants are male fertile which are removed after flower initiation through pollen examination. Remaining 50 per cent plants which are male sterile are allowed to be pollinated by male parent by open pollination. Hand pollination can also be practiced to achieve better hybrid seed setting. Other important points are as follows:

(i) Seed of parents Used: Foundation or Certified seed.

(ii) Planting Ratio of A and R Lines: 6 : 2 in adjacent rows or in 3: 1 ratio in adjacent blocks. Moreover, male parent is planted all around the seed production plot to ensure better seed setting and avoid contamination from other sources.

(iii) Isolation Distance: 200 meters.

(iv) Rogueing: Carefully done both in female and male parents to ensure high genetic purity of hybrid seed.

(v) Agronomic Practices: Recommended agronomic practices have to be followed.

(vi) Plant Protection Measures: Recommended plant protection measures are adopted.

(vii) Field Inspections: Four field inspections are made by seed inspectors of State Seed Certification Agency to check isolation distance, planting ratio, off types, disease and insect incidence and general crop conditions.

Q13. Explain briefly main points related to hybrid seed production in Sunflower.

Ans. Sunflower is a highly cross pollinated crop in which cross pollination occurs by wind and insects. Sunflower is grown in all seasons. Hybrid seed production is carried out through open pollination. However, for better seed setting hand pollination is also done. In sunflower, cytoplasmic genic male sterility is used for hybrid seed production. In India, about 80-85 per cent of total sunflower area is covered by hybrids. Other important points are as follows:

(i) Seed of parents Used: Foundation or Certified seed.

(ii) Planting Ratio of A and R Lines: 6 : 2.

(iii) Isolation Distance: 200 meters.

(iv) Rogueing: Carefully done both in female and male parents to ensure high genetic purity of hybrid seed.

(v) Agronomic Practices: Recommended agronomic practices have to be followed.

(vi) Plant Protection Measures: Recommended plant protection measures are adopted.

(vii) Field Inspections: Four field inspections are made by seed inspectors of State Seed Certification Agency to check isolation distance, planting ratio, off types, disease and insect incidence and general crop conditions.

Chapter 14

Wheat and Rice Breeding

Q1. Explain briefly about family, distribution and uses of Wheat.

Ans. It includes, family, global distribution, national distribution, types, uses as briefly presented below:

(i) Family: Wheat belongs to the family Poaceae (old *Gramineae*). It is an important cereal crop of cool climates.

(ii) Global Distribution: It is widely grown the world over and stands first among cereals in area and production. Wheat is cultivated in USA, U.K. Russia, Ukraine, China, Japan, Argentina, Canada, Mexico, India, Pakistan and many other countries.

(iii) National Distribution: In India, Uttar Pradesh, Madhya Pradesh, Punjab, Haryana, Rajasthan, Bihar and Gujarat are the major wheat growing states.

(iv) Types: Wheat is classified on various basis. Based on consumption it is classified as bread wheat, common wheat or soft wheat (T. aestivum) and hard wheat or macaroni wheat (T. durum). Based on cold tolerance it is of two types, *viz.* winter wheat (which can withstand chilling temperature) and spring wheat.

(v) Uses: Wheat is used for human consumption in a variety of products such as bread, cakes, noodles, cookies, chapati, macaroni *etc.*, in various countries.

Q2. Explain briefly about cultivated species, origin and evolution of Wheat.

Ans. There are two cultivated species of wheat, *viz.* common wheat (*Triticum aestivum*) and durum wheat (*T. turgidum* L.). Common wheat is hexaploid ($2n = 6x = 42$), whereas durum wheat is tetraploid ($2n = 4x = 28$). The former is more widely adapted than the latter. Common wheat is used for bread, cakes, noodles, cookies, chapati *etc.*, whereas the durum wheat is used mainly for macaroni and some flat bread. There are 16 wild species of wheat, out of which six are diploids ($2n = 14$), seven tetraploids ($2n = 4x = 28$), and three hexaploids ($2n = 6x = 42$). Wild species are used in hybridization programs for transfer of resistance to biotic and abiotic factors, adaptation and other desirable characters into cultivated species.

Origin and Evolution: Near East is the center of origin of bread wheat. Genetic investigations have established that bread wheat has evolved from crosses among three species of wheat in nature in two steps [belonging to A, B and D genome] as follows:

(a) Einkorn wheat [*T. monococcum* x unknown species] = Emmer wheat [*T. turgidum*]

(b) Emmer wheat x Einkorn wheat [*T. tauschi*] = bread wheat [*T. aestivum*]

Durum wheat has originated from a cross between two diploid wild species.

(i) Origin of tetraploid wheat [Triticum turgidum]

Parents	*Triticum monococcum*	x	Unknown species
Genome	AA [2n=14]		BB [2n=14]
F_1		AB [2n=14] Sterile	

Chromosome doubling in nature AABB [2n =28] *Triticum turgidum* [Tetraploid-Fertile]

(ii) Origin of hexaploid wheat [Triticum aestivum]

Parents	*Triticum turgidum*	x	*Triticum tauschi*
Genome	AA BB [2n=28]		DD [2n=14]
F_1		ABD [2n=21] Sterile	

Chromosome doubling in nature AABBDD [2n =42] *Triticum aestivum* [Hexaploid-Fertile]

FIGURE 14-1. Probable Origin of Hexaploid Bread Wheat [*Triticum aestivum*].

The chromosome doubling took place in nature in above crosses. Other two species of hexaploid wheat , *viz.* T. compactum, T. spherococcum originated through spontaneous mutation of T. aestivum. The durum wheat probably originated from cultivated Emmer wheat [T. turgidum var dicoccum] after several spontaneous mutations.

Q3. Explain briefly mode of reproduction and pollination in Wheat.

Ans. Wheat is a self-pollinated and seed propagated crop. Main points are presented as follows:

(i) Flowers: Flowers are bisexual. Flowers open generally after pollination is over (chasmogamy). This chamogampus protomes autogamy. The flower is covered by two membraneous structures. The outer is called and inner as palea.

(ii) Pollination: It is self-pollinated crop. Cross pollination is less than 1 per cent.

(iii) **Propagation:** Sexually produced seeds are used for propagation.

(iv) **Tillering:** Tillering is a common feature of wheat crop.

(v) **Isolation Distance:** An isolation distance of 3 metres is absolutely safe for the production of breeder and foundation seed.

Q4. What are Breeding Objectives in Wheat?

Ans. In wheat, major breeding objectives are higher grain yield, better quality, early maturity and wider adaptability and resistance to biotic (diseases and insects) and abiotic (drought, salinity, lodging, *etc.*) stresses.

(i) **Higher grain yield**: Major yield components are: ear length, kernel per ear, number of productive tillers per plant and kernel size.

(ii) **Quality:** Grain quality includes color, size and lusture, protein content, lysine content, *etc.*

(iii) **Disease Resistance:** Major diseases include rusts (black or stem rust, brown or leaf rust and yellow or stripe rust), loose smut, hill bunt, Karnal bunt, powdery mildew and leaf blight.

(iv) **Insect Resistance:** In India insects are not the serious problem. In USA, Hessian fly, stem sawfly and green bugs are major insect pests of wheat.

Q5. Explain briefly the breeding procedures used in Wheat.

Ans. Breeding procedures which are used for wheat improvement are of three types, *viz.* general methods, special methods and population improvement procedures as follows:

(i) **General Methods:** These methods include introduction, pure line selection, pedigree, back cross and multiline breeding, bulk and single seed descent methods.

(ii) **Special methods:** Such methods include: mutation breeding, distant hybridization, heterosis breeding and transgenic breeding.

(iii) **Population Improvement Procedures:** Such procedures include recurrent selection and diallel selective mating systems. These are used for specific purposes.

Q6. What are breeding centers of Wheat?

Ans. There are two types of wheat breeding centers as follows:

(i) **International Centers:** programs of wheat improvement are carried out by the There is only one International wheat breeding center, *i.e.* International maize and Wheat Improvement Centre (CIMMYT), Mexico. International multi-locational testing is carried out by CIMMYT for identification and release of varieties for different countries. The global gene poll of wheat is also maintained at this centre.

(ii) **National Centers:** In India, wheat breeding work is carried out by following organizations:

(a) Indian Institute of Wheat and Barley Research, Karnal. The new varieties are released through coordinated project after multi-location testing for 3-5 years.

(b) State Agricultural Universities located in wheat growing regions.

(c) Some private seed companies.

Q7. What are Practical Achievements of Wheat breeding?

Ans. Remarkable work on wheat breeding has been done by CIMMYT which led to green revolution in wheat production the world over. The semi-dwarf varieties of wheat have been developed through the use of Japanese line Norin 10 as a source of dwarfing gene. The semi dwarf varieties developed by CIMMYT at Mexico have spread to different wheat growing countries resulting in revolution in wheat production. The productivity of semi-dwarf varieties is about two and half times more than old tall growing varieties. Moreover, semi-dwarf varieties are highly resistant to lodging and are highly responsive to fertilizer doses. In India, several high yielding varieties of wheat have been released for different states through coordinated project. Some high yielding varieties of common wheat for different situations are listed below:

(i) Irrigated timely sown: HD 2329, HD 2009, HD 2428, PBW 154, CPAN 3004 (Punjab, Haryana, Western U.P., Rajasthan)

(ii) Irrigated late sown: HD 2285, HD 2270, PBW 226, PBW 138 (Above 4 states)

(iii) Rain-fed timely sown: C 306, WL 410, IWP 72, PBW 65, PBW 175, WL 2265 (Above States)

(iv) Salt affected soils: WH 157, KRL 1-4

(v) Hilly Regions: CPAN 1796, HD 208, HD 2380, HS 240, UP 1109 (U.P., J&K and H.P.) Varieties of durum wheat are PBW 34, PBW 215 and Raj 1555 for irrigated timely sown areas and JK 12 for rainfed timely sown areas.

Q8. Explain briefly about family, distribution and uses of Rice.

Ans. **(i) Family:** Rice belongs to the family Poaceae [earlier *Gramineae*]. It is an important cereal food crop of global significance. It It is grown in humid tropical and subtropical climates.

(ii) Global Distribution: The major rice growing countries are China, India, Japan, Pakistan, Southeast Asia and Brazil. Rice is one of the principal food crops of the world.

(iii National Distribution: In India, rice is grown in almost all the states.

(iv) Types: Rice is classified on the basis of fineness of grain as fine, and coarse; on the basis of scent as scented and non-scented.

(v) Uses: Rice is consumed as human food in a variety of ways, such as pullao, idle, plain rice after boiling, for making dosa and several other food preparations.

Q9. Explain briefly about cultivated species, origin and evolution Rice.

Ans. There are two cultivated species of rice, *viz.* Asian rice (*Oryza sativa* L.) and African rice (*Oryza glaberrima*). Both the species are diploid ($2n = 24$). Asian rice is the predominant species which has spread to different parts of the world. The African rice is still confined to tropical west Africa. There are three races of Asian rice, *viz.* Indica, Japonica and Javanica. Indica is grown in tropical climate, Japonica in the temperate region and the race Javanica is intermediate between Indica and Japonica. There are 22 wild species of rice which are found in the tropics of the world. Out of these, 8 are tetraploid species ($2n = 4x = 48$).

Origin and Evolution: Rice has originated in Asia as well as in Africa. The Asian rice [*Oryza sativa*] has originated from perennial Asian wild species [*Oryza rupipogan*] and the African cultivated rice [*Oryza glaberrema*] is believed to have originated in West Africa from wild diploid species [*Oryza breviligulata*]

Q10. Explain briefly mode of reproduction and pollination in Rice.

Ans. Rice is a self-pollinated and seed propagated crop. Main points are presented as follows:

(i) **Flowers:** Flowers are bisexual.

(ii) **Pollination**: Rice is self-pollinated crop. Cross pollination is less than 1 per cent.

(iii) **Propagation:** Sexually produced seeds are used for propagation.

(iv) **Tillering**: Tillering is a common feature of wheat crop.

(v) **Isolation Distance:** An isolation distance of 3 meters is absolutely safe for the production of breeder and foundation seed.

Q11. What are Rice breeding objectives?

Ans. In rice, major breeding objectives are higher grain yield, better quality, early maturity and wider adaptability and resistance to biotic stresses as follows:

(i) **Higher grain yield:** Major yield components are: panicle length, number of grains per panicle, number of productive tillers per plant and test weight.

(ii) **Grain quality:** Grain quality includes color, size and texture, protein content, lysine content, fragrance *etc.* Cooking qualities include kernel elongation, volume expansion, water absorption, gel consistency, etyc.

(iii) **Disease Resistance:** Major diseases include blast, bacterial blight, stem rot, brown spot, virus disease, *etc.*

(iv) **Insect Resistance:** Major insects include, stem borer, brown hopper, Gandhi bug and Gall fly

Q12. Explain briefly the breeding procedures used in Rice.

Ans. Breeding procedures which are used for rice improvement are of three types, *viz.* general methods, special methods and population improvement procedures as follows:

(i) **General Methods**: These methods include introduction, pure line selection, pedigree, back cross and multiline breeding, bulk and single seed descent methods.

(ii) **Special methods**: Such methods include: mutation breeding, tissue culture, heterosis breeding and transgenic breeding.

(iii) **Population Improvement Procedures**: Such procedures include recurrent selection and diallel selective mating systems. These are used for specific purposes.

Q13. What are breeding centers of Rice?

Ans. There are two types of rice breeding centers as follows:

(i) **International Centers:** There is only one International rice breeding center, *i.e.* International Rice Research Institute, Philippines. International multi-locational testing is carried out by IRRI for identification and release of varieties for different countries. The global gene poll of rice is also maintained at this centre.

(ii) **National Centers:** In India, rice breeding work is carried out by following organizations:

(a) National Rice Research Institute, Cuttack, Odisha.

(b) Indian Institute of Rice Research, Hyderabad, Tenangana.

(c) State Agricultural Universities located in rice growing regions.

(d) Some private seed companies.

Q14. What are Practical Achievements of Rice breeding?

Ans. IRRI has played significant role in the genetic improvement of rice. Several semi-dwarf, high yielding, varieties with wider adaptation have been released by IRRI for different countries. The Chinese Indica line Dee Geo-woo-gen has been used as a source of dwarfing gene in evolving semi-dwarf varieties of rice. Evolution of semi-dwarf varieties resulted in green revolution in rice. Important rice varieties developed by IRRI include, IR 8, IR 20, IR 22, IR 24, IR 26, IR 28, IR 29, IR 30, IR 32, IR 34 and IR 36. Out of these, IR 8, IR 20 and IR 36 have been released for India. In China, hybrid rice using cytoplasmic genetic male sterility system has been developed for commercial cultivation. Hybrid rice has also been developed in India by Directorate of Rice Research, Hyderabad. However, its commercialization will take some more time. In India, more than 100 improved varieties of rice for different states have been released by CRRI, DRR and State Agricultural Universities. The bacterial blight resistant varieties developed by IRRI and salinity resistant, scented and high yielding varieties developed for cultivation in India are listed below:

1. **Bacterial blight resistant varieties:** IR 20, IR 22, IR 26, IR 28, IR 29, IR 30, IR 32 and IR 34.
2. **Salinity resistant varieties**: Usar Dhan 1 (U.P.), PVR 1 (Tamil Nadu), MCM 1, MCM 2 (A.P.) Bhurarata, Kalarata (Maharashtra), Korekagga, Bilikagga (Karnataka), Damodar, Getu, Dosal, Nanobokhra, Nonasail, Rupsail (West Bengal), Pokkai and Chotupokkali (Kerala).
3. **Scented varieties**: Type 3, N 12, N 10 (B), Kasturi, Pusa, Basmati.
4. **High yielding varieties:** IR 8, IR 20, IR 36, Vikas, Savitri, Anamika, Mansarovar *etc.*

Chapter 15

Maize, Sorghum and Pearl Millet Breeding

Q1. Explain briefly about family, distribution and uses of Maize.

Ans. (i) Family: Maize belongs to the monocot family *Gramineae* (*Poaceae*).

(ii) Global Distribution: Maize is grown in Central and South America, Africa, Central Europe and Asia.

(iii) National Distribution: In India, U.P., Bihar, Rajasthan, M.P., Punjab and H.P. are the major maize growing states.

(iv) Types: Maize is of two types, *viz.* flints (with round seeds) and dents (with flat seeds). The flour can be easily made from flints than from dents. Maize has C4 photosynthetic pathway and is more efficient than C3 plants.

(v) Uses: Maize is an important cereal crop of global importance, which is used for human consumption in developing and undeveloped countries and as livestock feed in developed countries. Moreover, it is also used as green fodder for livestock.

Q2. Explain briefly about cultivated species, origin and evolution of Maize.

Ans. There are five species of the genus Zea *viz. Zea maxicana, Z. perennis, Z. luxurian, Z. diploperenis* and *Z. mays*. The first four species are wild and commonly called *Teosinte*. The last species *i.e, Z. mays* is the only cultivated species. The teosinte species are easily crossed with maize. The diploid chromosome number in maize is 20. The wild species, *Z. perennis* is tetraploid ($4n = 40$); rest are diploids. The other genus which is close to maize is *Tripsacum*. However, special techniques are required to cross maize and *Tripsacum*. There are 16 species of *Tripsacum* in which the somatic chromosome number varies from 36 to 108. The tetraploid wild species *Z. perennis* is believed to have originated from *Z. mexicana*.

Origin and Evolution: Central America (Mexico) is the place of origin of maize. The exact genetic origin of maize is not known so far. There are four different views (hypotheses) about the possible genetic origin of maize which are summarized below:

1. Maize, *Teosinte* and *Tripsacum* have originated from a common ancestor (from high lands of Mexico and Guatemala) which is now extinct (Weather wax, 1955).
2. Maizeoriginated probably from a cross between Coix and Sorghum each with 10 chromosomes (Anderson, 1945).
3. Modern maize is believed to have originated from crosses between wild maize (pod corn) and Tripsacum. The Teosinte probably developed from a cross between cultivated maize and Tripsacum in Central America (Mangelsdorf and Reeves, 1939).
4. Some workers believe that maize has derived fro Teosinte by direct selection (Beadle, 1939).

Q3. Explain briefly about mode of reproduction and pollination in Maize.

Ans. Maize is a cross-pollinated and seed propagated crop. Main points are presented as follows:

(i) Flowers: Maize is a monoecious plant. Male and female flowers are separate on the same plant. Male inflorescence is called tassel and female flower is known as silk. Tassel is present on the top and silk in the middle portion of the plant.

(ii) Cross Pollination: Under natural condition, it is more than 99 per cent. Pollen is very light and cross pollination occurs by wind.

(iii) Dichogamy: In maize male and female flowers mature at different time which promotes cross pollination.

(iv) Propagation: Maize is a seed propagated crop.

(v) Tillering: Maize is a non-tillering cereal.

(vi) Isolation Distance: An isolation distance of 400 m and 200 m is required for the production of foundation seed of open pollinated varieties.

Q4. What are Breeding Objectives of Maize breeding?

Ans. In maize, major breeding objectives are higher grain yield, better quality, early maturity and wider adaptability and resistance to biotic (diseases and insects) and abiotic (drought, salinity, lodging, *etc.*) stresses as follows:

(i) Higher grain yield: Major Yield Components are: ear length, kernel per row and per ear, size of kernel and number of ear per plant.

(ii) Grain Quality: Grain quality includes grain color, protein content, lysine and tryptophan. The gene opaque 2 is the source of high lysine content. The protein found in endosperm is known as Zein and is deficient in lysine and tryptophan.

(iii) **Disease Resistance**: Major diseases are root rot, stalk and ear rot, leaf blight, leaf spot, downy mildew and mosaic virus.

(iv) **Insect Resistance**: In India stem borer is the major insect pest.

Q5. Explain briefly the breeding procedures used in Maize.

Ans. Breeding approaches which are used for maize improvement are of three types, *viz.* population improvement, inbred development and cultivar development.

(i) **Population Improvement Methods**: These methods include mass selection, progeny selection or ear to row selection, recurrent selection, disruptive mating and selection and bi-parental mating.

(ii) **Inbred Improvement Methods**: Such procedures include back crossing, convergent improvement and gamete selection. These are used for improving inbred lines for various economic traits,

(iii) **Methods for Cultivar Development**: Mass selection is used for development of open pollinated cultivars and their maintenance. Heterosis breeding is used for developing high yielding hybrids. In maize three types of hybrids are developed, *viz.* single cross hybrids, three way cross hybrids and double cross hybrids. Synthetic and composite varieties are also developed to partially exploit heterosis. Tissue culture helps in developing haploids and creating additional variability.

Q6. What are breeding centers of Maize?

Ans. There are two types of maize breeding centers as follows:

(i) **International Centers:** There is only one International wheat breeding center, *i.e.* International maize and Wheat Improvement Centre (CIMMYT), Mexico. International multi-locational testing is carried out by CIMMYT for identification and release of varieties for different countries. The global gene poll of wheat is also maintained at this center.

(ii) **National Centers:** In India, maize breeding work is carried out by following organizations:

(iii) Indian Institute of Maize Research, Ludhiana. The new varieties are released through coordinated project after multi-location testing for 3-5 years.

(iv) Vivekanand Parvatiya Krishi Anusandhan Shala, Almora, especially for hill regions.

(v) State Agricultural Universities located in maize growing regions.

(vi) Some private seed companies.

Q7. What are Practical Achievements of Maize breeding?

Ans. In maize, several composite and hybrid cultivars have been released by AICMIP for various agro-climatic conditions of the country. Some important high yielding composite and hybrid cultivars with wider adaptability are listed below:

Composite: Jawahar, Vijay, Kisan, Amber, Sona, Vikram, Protina, Rattan, Shakti, Vikas, Navjot, Ageti 76, Kanchan, Diara 3, D 765, MCU 508, Kiran, Surya, Tarun, Arun, Renuka, *etc.*

Hybrids: Ganga safed, Ganga 5, Ganga 9, Deccan 103, Sartaj Makka 1, VL 42 *etc.* Hybrids Him 123 and Him 128 have been released for the hilly regions of UP, HP, Assam and Sikkim. Both genetic and cytoplasmic genetic male sterility systems are available in maize. Usually, CGMS system is used for the development of single, three-way and double cross hybrids in maize.

Q8. Explain briefly about family, distribution and uses of Sorghum.

Ans. (i) **Family:** *Sorghum* belongs to the family *Gramineae*. It is an important food crop of global significance which is cultivated in tropical and subtropical climates especially in the semiarid tropics the world over. *Sorghum* is a C4 plant and, therefore, has high photosynthetic efficiency.

(ii) **Global Distribution**: The major *Sorghum* producing countries are Africa, India, China, Pakistan, USA, Australia, Argentina and Mexico.

(iii) **National Distribution**: In India, major *Sorghum* growing states are Maharashtra, Gujarat, Tamil Nadu, Karnataka, Rajasthan and Madhya Pradesh.

(iv) **Types**: It is of two typed, *viz.* grain sorghum and fodder sorghum.

(v) **Uses:** The grain is used as human food in various ways and stalk as animal feed. In south India, where winter is mild, it is also grown in Rabi season.

Q9. Explain briefly about cultivated species, origin and evolution of Sorghum.

Ans. There is only one cultivated species of *Sorghum*, *i.e. Sorghum bicolour* which is diploid ($2n = 2x = 20$). In *Sorghum*, the panicle has wide variability ranging from very open and loose to compact. Based on panicle morphology, there are five races of *Sorghum*, *viz.* Bicolour (open panicle and elongated grains), Guinea (open panicle and flattened grains), Caudatum (high yield and bright seed color), Kafir (Semi compact to compact panicle and spherical grain), and Durra (round grains). The races Kafir and Durra have several useful genes. Stem juice of fodder *Sorghum* is sweeter than grain *Sorghum*. The important wild species are *S. bicolour* spp. *arundinaceum* ($2n = 20$) and *S. halepense* ($2n = 4x = 40$).

Origin and Evolution: It is generally accepted that Ethopia is the centre of origin of cultivated *Sorghum*. From there it spread to other parts of the world. The diploid wild species *S. arundinaceum* ($2n = 20$) is considered to be the likely ancestor of cultivated *Sorghum*. This wild species is still found growing in Ethopia and its adjacent regions.

Q10. Explain briefly about mode of reproduction and pollination in Sorghum.

Ans. The cultivated *Sorghum* is seed propagated and self-pollinated. Main points are presented as follows:

(i) **Pollination**: It ia a self-pollinated crop. Natural cross pollination varies from 5 to 30 per cent depending upon genotype and environmental conditions. Pollen is very light and cross pollination occurs by wind.

(ii) **Protogyny**: It is common which promotes cross pollination.

(iii) **Propagation**: Sorghum is a seed propagated crop.

(iv) **Tillering**: In Sorghum, tillering is important.

(v) **Isolation Distance**: An isolation distance of 200 m. and 100 m. is essential for the production of foundation and certified seed.

Q11. What are Sorghum breeding Objectives?

Ans. In Sorghum, major breeding objectives are higher grain yield, better quality, early maturity and wider adaptability and resistance to biotic (diseases and insects) and abiotic (drought and lodging) stresses as follows:

(i) **Higher grain yield**: Major Yield Components are: panicle size and type, grains per panicle, productive tillers per plant and test weight.

(ii) **Quality**: Quality characters include chapatti and bread making qualities, protein and lysine content.

(iii) **Disease Resistance**: Major diseases are leaf blight, zonate leaf spot, downy mildew, rust, ergot or sugary disease and charcoal rot.

(iv) **Insect Resistance**: In India stem borer, stem fly and midge are important diseases. Striga is a root parasite weed in Sorghum.

Q12. Explain briefly the breeding procedures used in Sorghum.

Ans. Breeding approaches which are used for Sorghum improvement are of two types as presented below:

(i) **General Methods**: These methods include introduction, pure line selection, pedigree breeding, back crossing and composite breeding.

(ii) **Special Methods**: Such methods include, mutation breeding disruptive mating and selection and bi-parental mating. Biotechnology is expected to play significant role in in future in genetic improvement of Sorghum.

Q13. What are Sorghum breeding centers?

Ans. There are two types of maize breeding centers as follows:

(i) **International Centers:** At International level Sorghum improvement work is carried out by International Crop Research Institute for Semi-arid Tropics (ICRISAT), Hyderabad (India). New varieties are developed after multi-location testing in different countries. Testing centers are located in Nigeria, Kenya, Zimbabve, Mexico and India.

(ii) National Centers: In India, Sorghum breeding work is carried out by following organizations:

(a) Indian Institute of Millet Research, Hyderabad. The new varieties are released through coordinated project after multi-location testing for 3-5 years.

(b) State Agricultural Universities located in Sorghum growing regions.

(c) Some private seed companies.

Q14. What are Practical Achievements of Sorghum breeding?

Ans. In India, several high yielding varieties and hybrids have been developed for different states. Hybrids are developed by using cytoplasmic genic male sterility. Kafir 60 is the sourceof male sterility. High lysine Ethiopian mutants (IS 11167 and 11758) are used for quality improvement. Some important high yielding varieties and hybrids of grain Sorghum are listed as follows:

1. **Hybrids:** CSH 1, CSH 2, CSH 3, CSH 4, CSH 5, CSH 6, CSH 7, CSH 8, CSH 9, CSH 10 and CSH 11 for Kharif and , CSH 7, CSH 8, CSH 12, and CSH 13 for Rabi season.
2. **Varieties:** CSV 1, CSV 2, CSV 3, CSV 4, CSV 5, CSV 6, CSV 7, CSV 8, CSV 9, CSV 10 and CSV 11, CSV 12, CSV 13, M 35-1 and Swati for Kharif and , CSV 7 and CSV 8, for Rabi season.

Q15. Explain briefly about family, distribution and uses of pearl millet.

Ans. **(i) Family**: Pearl millet belongs to the family Poaceae (old *Gramineae*). It is an important food crop of semi-arid tropics.

(ii) Global Distribution: Major pearl millet growing countries are Africa, India, Pakistan and South East Asia. In USA and Europe, pearl millet is known as fodder crop.

(iii) National Distribution: In India, major pearl millet growing states Rajasthan, Gujarat, Maharashtra, Uttar Pradesh and Haryana.

(iv) Uses: Grains are used as food and stalk as cattle fodder.

Q16. Explain briefly about cultivated species, origin and evolution of Pearl millet.

Ans. There is only one cultivated species of pearl millet, *i e. Pennisetum americanum*. Earlier it was known as *P. typhoides*. The cultivated species is diploid ($2n = 14$). There are over 50 wild species of pearl millet in which chromosome number is multiple of 5, 7, 8 and 9. Two wild species, *viz.* Napier grass (*P. purpureum*) and *P. squamulatum* have been used in breeding programs to develop perennial fodder varieties, and drought and cold resistant cultivars, respectively.

Origin and Evolution: Vast genetic diversity in pearl millet is found in West Africa and, therefore, it is generally accepted that Western Africa is the original home of pearl millet. It spread to other countries from Africa in course of time. Nothing is so far known about the ancestor of cultivated pearl millet.

Q17. Explain briefly mode of reproduction and pollination in pearl millet.

Ans. (i) **Pollination**: Pearl millet is a cross-pollinated crop. In this crop natural cross pollination is more than 80 per cent. Pollen is very light and cross pollination occurs by wind.

(ii) **Protogyny**: It is common which promotes cross pollination.

(iii) **Propagation**: Pearl millet is a seed propagated crop. Some wild species of pearl millet such as Napier grass propagates asexually by means of rhizomes.

(iv) **Tillering**: In pearl millet, tillering is important.

(v) **Isolation Distance**: An isolation distance of 400 m. and 200 m is essential for the production of foundation and certified seed.

Q18. What are Breeding Objectives in pearl millet?

Ans. In pearl millet, major breeding objectives are: higher grain yield, better quality, early maturity and wider adaptability and resistance to biotic (diseases and insects) and abiotic (drought and lodging) stresses as follows:

(i) **Higher grain yield**: Major Yield Components are: ear size and compactness, number of productive tillers per plant and test weight.

(ii) **Grain quality**: Bold, lustrous and pearly amber grains are considered desirable for human consumption in the form of various preparations.

(iii) **Disease Resistance**: Major diseases are Downy mildew, ergot and rust.

(iv) **Insect Resistance**: In India, Insects are not serious problems in this crop.

Q19. Explain briefly the breeding procedures used in pearl millet.

Ans. Breeding approaches which are used for pearl millet improvement are of two types as presented below:

(i) **General Methods**: These methods include introduction, mass selection, back cross method, heterosis breeding, synthetic and recurrent selection. disruptive mating and selection, bi-parental mating.

(ii) **Special Methods**: Such methods include, tissue culture and biotechnology which are expected to play significant role in in future in genetic improvement of pearl millet.

Q20. What are breeding centers of pearl millet?

Ans. There are two types of pearl millet breeding centers as follows:

(i) **International Centers:** At International level pearl millet improvement work is carried out by International Crop Research Institute for Semi-arid Tropics (ICRISAT), Hyderabad (India). New varieties are developed after multi-location testing in different countries. Testing centres are located in Nigeria, Kenya, Zimbabve, Mexico and India. ICRISAT maintains global gene pool of pearl millet. ICRISAT deals with five crops *viz.* pearl millet, *Sorghum*, chickpea, pigeon pea and groundnut.

(ii) National Centers: At National level, pearl millet breeding work is carried out by following organizations:

(a) Central Arid-zone Research Institute, Jodhpur The new varieties are released through coordinated project after multi-location testing for 3-5 years. All India coordinate pearl millet improvement project is located at Agricultural Research Station, RAU, Mandore, Jodhpur (Rajasthan).

(a) State Agricultural Universities located in pearl millet growing regions.

(b) Some private seed companies.

Q21. What are Practical Achievements of pearl millet breeding?

Ans. Several high yielding varieties and hybrids of pearl millet have been developed in India for different pearl millet growing states. Two varieties, *viz.* ICMS 7703 and ICTP 8203 and two hybrids, *viz.* ICMH 423 and ICMH 451 have been released by ICRISAT for India. Important varieties developed by Research Stations of Agricultural Universities and other organizations are listed below:

(i) Hybrids: HHB 45, HHB 50, HHB 60, HHB 67 from Hisar; GHB 27, GHB 30, GHB 181 FROM Jamnagar [Gujarat]; Pusa 3 from IARI, New Delhi; PHB 17 from Punjab; Other hybrids include HBH 128, MBH 130, MH 143, MH 169, MH 179, MH 180, MH 82 MH 208, *etc.* All hybrids are CMS based. Tift 23 A and Ludhiana cytoplasm have been used as sources of CMS in pearl millet improvement.

(ii) Varieties: PSB 8, PSB 15, PSB 2, HC 4, RHR 1, PCB 15, Mukta, Mallikarjun, Pusa safed Bajra, *etc.*

Chapter 16

Breeding Commercial Crops

Q1. Explain briefly about family, distribution and uses of Cotton.

Ans. Important commercial field crops include cotton, sugarcane, tobacco and potato.

(i) **Family:** Cotton belongs to the family *Malvaceae* and produce spinnable seed coat fibers. There are two types of fibres, *viz.* long fibres called lint and small fibres called linters or fuzz.

(ii) **Global Distribution**: Cotton is grown in tropical and subtropical regions of more than 80 countries of the world. The major cotton growing countries are China, USA, India, Pakistan, Uzbekistan, Egypt, Sudan, Turkey *etc.*

(iii) **National Distribution**: In India, there are nine major cotton growing states, *viz.* Punjab, Haryana, Rajasthan (North Zone), Maharashtra, Madhya Pradesh, Gujarat (Central Zone), Andhra Pradesh, Karnataka and Tamil Nadu (South Zone).

(iv) **Types**: Cotton is of two typed as Old World cottons and New World cotton. Cotton is also classified as diploid cotton and tetraploid cotton.

(v) **Uses**: Cotton is a fibre, oil and protein yielding crop. The lint is used in the textile mills, linters for various industrial uses, refined cotton seed oil is used for human consumption and the cake as fertilizer and livestock feed.

Q2. Explain briefly about cultivated species, origin and evolution of Cotton.

Ans. In the genus gossypiun about 50 species have been identified so far. Out of these four species, *viz., Gossypium hirsutum, G. barbadense, G. arboreum* and *G. herbaceum* are cultivated and rest are wild or uncultivated. The first two species are tetraploid [2n =52] and last two species are diploid [2n =26]. Tetraploid cottons are also known as New World cottons and diploid cottons are also known as Old World cottons or Asiatic cottons. India is the only country where all the four cultivated species are commercially planted. The *G. hirsutum* is also known as upland cotton or American cotton. *G. barbadense* is also known as Egyptian cotton or Peruvian cotton or Tanguish cotton or Sea Island cotton.

Origin and Evolution: Central Africa is considered as the centre of origin of the genus *Gossypium,* where *G. africanum* is still found growing in the wild state. Genetic origin of four cultivated species is discussed as follows:

(i) Diploid Cottons: It is believed that diploid species of cotton originated first and tetraplois species later on form a cross between two diploid species. The wild linted species *Gossypium africanum* is considered as the progenitor of all cultivated species of cotton. *Gossypium africanum* reached undivided India from South Africa through travelers, traders and explorers and after differentiation gave birth to two diploid cultivated species, *viz. Gossypium arboretum, Gossypium herbaceum.*

(ii) Tetraploid Cottons: There are two species of tetraploid cultivated cotton, *viz. Gossypium hirsutum* and *G. barbadense*. The genome constitution of both these species is AD. It is believed that A genome of tetraploid cotton has come from Asian cultivated species and D genome from American lintless wild diploid species (Skovsted, 1934). Skovsted (1934) first pointed out that the New World tetraploid consisted of A and D sub-genomes. Later on origin of tetraploid cotton was explanined by other workers. There are two views about the genetic origin of tetraploid cotton as discussed below:

(1) Beasley and Harland View: In 1940, Beasley and Harland independently observed that A genome of upland cotton has come from Asian cultivated diploid species *Gossypium arboretum* and D genome from American wild lintless diploid species *G. thurberi*. The cross between these two species took place in nature followed by chromosomal doubling. This is represented as follows:

Parents	*Gossypium arboretum*	x	*Gossypium thurberi*
Genome	AA [2n=26, large]		DD [2n =26, small
F1		AD [Sterile]	
Chromosome doubling		AA DD Fertile – Like *G. hirsutum*	
		[4n+ 52, 26 large and 26 small]	

FIGURE 16-1. Probable Origin of *G. hirsutum* as Proposed Independently by Beasley (1940) and Harland (1940).

(2) Phillips (1963) View: He reported that A genome of upland cotton has come from African linted wild diploid species *Gossypium africanum* and D genome from American wild lintless diploid species *G. raymondii.* The cross between these two species took place in nature followed by chromosomal doubling. This is represented as follows:

Parents	*Gossypium africanum*	x	*Gossypium raymondii*
Genome	AA [2n=26, large]		DD [2n =26, small
F1		AD [Sterile]	
Chromosome doubling		AA DD Fertile – Like *G. hirsutum* [4n+ 52, 26 large and 26 small]	

FIGURE 16-2. Probable Origin of *G. hirsutum* as proposed Phillips (1963).

Cytological, biochemical [electrophoretic studies] and molecular investigations have clearly indicated that A genome of *G. africanum* is closer to A genome of tetraploid cotton than that of *G. arboretum*. Similarly, D genome of *G. raymondii* is closer to D genome of tetraploid cotton than that of *G. thurberi*. Thus *G. africanum* and *G. raymondii* are progenitors of tetraploid cotton. The second view is widely accepted.

Hutchinson, Silow and Stephens (1947) presented classification of the genus Gossypiun on the basis of cytological, genetic, geographical and archeological evidences. They divided all the cultivated and wild species of cotton in to eight sections, *viz.*, Sturtiana, Erioxyla, Klotzschiana, Thurberana, Anomala, Stockiana, Herbacea and Hirsuta. Later on Phillips added one more section *i.e.* Longicalyciana.

Q3. Explain briefly about reproduction and pollination of Cotton.

Ans. Cotton is a self-pollinated and seed propagated crop. Main points are presented as follows:

(i) **Flowers**: Flowers are bisexual. Flowers are showy and attracts insects.

(ii) **Pollination**: Cotton is basically a self-pollinated crop. Cross pollination varies from nil to 60 per cent under different climatic conditions.

(iii) **Apomixis**: In cotton semi-gamy is reported

(iv) **Propagation**: Sexually produced seeds are used for propagation.

(v) **Branching**: Cotton crop has two types of branches, *viz.* monopodia and sympodia.

(vi) **Isolation Distance**: An isolation distance of 50 m and 30 m is required for the foundation of and certified seed.

Q4. What are Breeding Objectives of Cotton?

Ans. In cotton, major breeding objectives are higher yield, better fibre quality, early maturity, wider adaptability and resistance to biotic and abiotic stresses as follows:

(i) **Higher Lint Yield**: Major yield components are: number of bolls per plant and boll weight.

(ii) **Fibre Quality**: Fibre qualities include fibre length, fibre strength, fibre fineness, fibre maturity, *etc.*

(iii) **Disease Resistance**: Major diseases include Fusarium wilt, Vertyicillium wilt, bacterial

(iv) **Insect Resistance**: Important insects include boll worms, jassids, aphids, white fly, thrips, mites, *etc.*

Q5. Explain briefly the breeding procedures used in Cotton.

Ans. Breeding procedures which are used for cotton improvement are of three types, *viz.* general methods, special methods and population improvement procedures as follows:

(i) **General Methods**: These methods include introduction, pure line selection, pedigree, back cross and multiline breeding, bulk and single seed descent methods.

(ii) **Special methods**: Such methods include: mutation breeding, distant hybridization, heterosis breeding and transgenic breeding.

(iii) **Population Improvement Procedures**: Such procedures include recurrent selection and diallel selective mating systems. These are used for specific purposes.

Biotechnology has helped in developing transgenic cotton with resistance to *Helicoverpa.* The resistant gene has been transferred from bacteria *Bascillus thuringiensis* into cotton plant by Mansanto Seed Company in USA.

Q6. What are breeding centers of Cotton?

Ans. There are two types of cotton breeding centers as follows:

(i) **International Centers:** No international research centre for cotton breeding has been established so far. There is International Cotton Advisory Committee (ICAC) which advises on various matters of cotton research and marketing.

(ii) **National Centers:** In India, cotton breeding work is carried out by following organizations:

(a) Central Institutes for Cotton Research, Nagpur. The new varieties are released through coordinated project after multi-location testing for 3-5 years.

(b) State Agricultural Universities located in cotton growing regions.

(c) Some private seed companies.

Q7. What are Practical Achievements of cotton breeding?

Ans. There are three main achievements in the improvement of cotton crop in India after independence. These are: (1) acclimatization of *barbadense* cotton, (2) improvement of *hirsutum* and (3) development of hybrids. Egyptian cotton got acclimatized in India in 1955. The yield of *hirsutum* varieties has been significantly improved. India is the pioneer country in the world for the commercial cultivation of hybrid cotton. Several cotton hybrids have been developed in India. However, most of them have been evolved by conventional method *i.e.* hand emasculation and pollination. One hybrid *viz.* Suguna has been developed through the use of genetic male sterility and two (PKVHY 3 and PKVHY 4) through the use of cytoplasmic genic male sterility so far. *G. hirsutum* and *G. arboreum* are cultivated in all cotton growing states. However, *G. barbadense* is cultivated in few pockets of Tamil Nadu and Andhra Pradesh, and *G. herbaceum* in some parts of Gujarat and Karnataka. The important varieties and hybrids of cotton are listed below:

(1) **Varieties:** *G. hirsutum.* LH 900, LH 1556, F 846, F 1378, F 1861, HS 6, H 874, H 1098, HS 182, H 1117 RST 9, RS 875, 810 (North zone), Khandwa 3, Vikram, PKV 081, Rajat, LRA 5166, LRK 516 (Central zone), Abadita, Sharada, Sahana, MCU 12, MCU 5 VT, MCU 10, and LRA 5166. (South zone) *G. arboreum.* LD 491, LD 694, RG 18, HD 107, HD 123, LD 327, RG 8 (North zone), AKH 4, AKA 8401, Maljari, AKA 5, AKA 7, PA 255, PA 402 (Central zone) K 9, K 10, K 11 (South zone).

(2) **Hybrids:** (*a*) **Interspecific hybrids.** Varalaxmi, DCH 32, DHB 105, Sruthi TCHB 213, NHB 12, HB 224 *etc.*

(*b*) **Intraspecific hybrids.** HH : H 4, H 6, H 8, H 10, NHH 44, JKHY 1, JKHY 2, PKV HY 2, PKVHY 3, PKVHY 4, PKVHY 5, Savita, CICR HH 1, Surya, DHH 11, Fateh, LHH 144, Dhanlaxmi, Maruvikas, Omshankar, CSHH 198, *etc.*

(*c*) **Desi hybrids.** DH 7, DH 9, DDH 2, Pha 46, LDH 11, AKDH 7, RAJDH 7, GCot MDH 11, AAH 1, CICR 2 *etc.*

(*d*) **Male Sterility based hybrids.** Suguna, PKVHY 3, PKVHY 4, PKVHY 4, AAH 1, ARDH 7, RAJDH7, G.Cot MDH 11, CICR 2.

(*e*) **Bt. Hybrids.** About 619 Bt. cotton hybrids have been released by different seed companies.

Q8. Explain briefly about family, distribution and uses of Sugarcane.

Ans. **(i) Family:** Sugarcane belongs to the family *Gramineae*. It is an important commercial crop which is grown in tropical and subtropical climates the world over. Being C4 plant, it has high photosynthetic efficiency.

(ii) Global Distribution: The important sugarcane growing countries are India, Brazil, Cuba, China, USA, Mexico, France, Germany and Australia.

(iii) National Distribution: In India, the major sugarcane growing states are Uttar Pradesh, Maharashtra, Andhra Pradesh, Tamil Nadu, Karnataka, Bihar, Punjab and Haryana.

(iv) Uses: Sucrose is the main product of sugarcane. The other products are green tops, bagasse, molasses, filter mud, stubbles, and dry leaves. Green tops are used as cattle feed; bagasse for production of paper, rayon grade pulp, particle boards, fibre boards, as a fuel and also for production of alcohol; molasses for production of citric acid, butamol, acetone, rum, glycol, alcohol and acetic acid; filter mud for production of crude wax, fatty lipids, waxy lipids and refined wax; stubbles for making compost and as fuel; and dry leaves as mulch, fuel and compost.

Q9. Explain briefly about cultivated species, origin and evolution of Sugarcane.

Ans. There are three cultivated species of sugarcane, *viz. Saccharum officinarum* ($2n = 80$), *S. sinense* ($2n = 118$) and *S. barberi* ($2n = 82\text{-}124$). The species *S. officinarum* is also known as noble cane. There are two wild species of sugarcane, *viz. S. spontaneum* and *S. robustum*. Main features of wild and cultivated species are as follows:

TABLE 18-1. Main Features of Wild and Cultivated Species of Sugarcane

Sl.No.	Species	Distribution	Main Characters
1	*Saccharum officinarum*	Grown in South India	Canes are thick, susceptible to drought, red rot and mosaic. Soft rind, low fibre, high sugar content, *etc.*
2	*S. barberi*	Grown in North India	Canes are thin with hard rind, early maturing, wider adaptability, low sugar content and resistance to red rot butsusceptible to smut.
3	*S. sinense*	Grown in North India	Canes are thin with hard rind, early maturing, wider adaptability, low sugar content and resistance to red rot but susceptible to smut.
4	*S. spontaneum*	Found in India, South and South East Asia,	Canes are thin with hard rind. This has high tillering capacity and wider adaptability. early maturing, wider adaptability, low sugar content and resistance to red rot but susceptible to smut.
5	*S. robustum*	Found in New Guinea	Wider adaptability, very thick canes, low sugar content, high fibre and susceptible to mosaic.

Origin and Evolution: It is believed that *Saccharum officinarum* has originated in New Guinea and *S. barberi* and *S. sinensis* in North India. Wild species *S. robustum* is considered to be the progenitor of cultivated species S. officinarum. In other words, noble cane originated in New Guinea and migrated to North India. After hybridization with wild species *S. spontaneum*, it gave birth to *S. barberi* and *S. sinense*. The overall process is presented as follows:

1. *S. robustum* — *S. officinarum* — New Guinea
2. *S. officinarum* x *S. spontaneum* — *S. barberi* and *S. sinense* — North India.

Q10. Explain briefly about reproduction and pollination of Sugarcane.

Ans. Sugarcane is a cross-pollinated and asexually propagated crop. Main points are presented as follows:

(i) Flowers: Flowers are bisexual and are called as arrow. Flowering requires warm nights, humid conditions and high rain fall. These conditions are available at Coimbatore.

(ii) **Pollination**: Sugarcane is a cross-pollinated crop.

(iii) **Apomixis**: In sugarcane apomixis is not reported.

(iv) **Propagation**: Sugarcane is asexually propagated.

(v) **Tillering**: Tillering is important in this crop.

(vi **Isolation Distance**: An isolation distance of 50 m and 30 m is required for the foundation of and certified seed.

Q11. What are Breeding Objectives of Sugarcane?

Ans. In sugarcane, major breeding objectives are higher cane yield, better fibre quality, early maturity, wider adaptability and resistance to biotic and abiotic stresses as follows:

(i) **Higher cane yield**: Major yield components are: stem height, thickness, tillering capacity, number of millable canes per plant and weight of individual cane.

(ii) **Quality**: Milling quality includes soft rind, long internode, less fibre, less pith, high sucrose content and thick juice.

(iii) **Disease Resistance**: Major diseases include red rot , smut,

(iv) **Insect Resistance**: Important insects include top borer, pyrilla, white fly, early shoot borer, and internode borer.

Q12. Explain briefly the breeding procedures used in Sugarcane.

Ans. Breeding methods which are used for the genetic improvement of sugarcane can be divided into three groups, *viz.* general methods, special methods and population improvement procedures as follows:.

(i) **General Methods:** These methods include introduction, clonal selection and backcrossing.

(ii) **Special Methods:** These methods include interspecific hybridization, intergeneric hybridization and mutation breeding. Distant hybridization (inter specific and intergeneric crosses) is used for transfer of resistance to biotic and abiotic stresses from wild species or allied genera into the cultivated species. Mutation breeding is mainly used for disease resistance.

(iii) **Population Improvement**: In sugarcane, population improvement program is also taken up. Bi-parental mating, disruptive mating and recurrent selection are used for population improvement. Tissue culture technique is used for development of haploids (anther-culture), earliness and for resistance to biotic and abiotic factors. Somaclonal variation is expected to help in evolving varieties resistant to biotic and abiotic stresses.

Q13. What are breeding centers of Sugarcane?

Ans. Sugarcane breeding work is carried out in each country at national research centre. No International Research Institute has been established for sugarcane breeding so far. In India, sugarcane breeding work is carried out by the following research Institutes/organizations :

1. Sugarcane Breeding Institute, Coimbatore.
2. Indian Institute of Sugarcane Research, Lucknow.
3. State Sugarcane Research Stations, such as Shahjahanpur (UP), Seorahi, Deoria (UP), Pusa (Bihar), Padegaon, (Maharashtra) and Anakapalli (AP). New varieties are released by All India coordinated Sugarcane Improvement Project after multi-location testing for 3-5 years. Coordinated Project is located at Lucknow in the premises of ISSR.

Q14. What are Practical Achievements of Sugarcane breeding?

Ans. Several improved varieties of sugarcane suitable for different states have been released by Sugarcane Breeding Institute, Coimbatore and State Sugarcane Research Stations through Coordinated Project. Some currently cultivated sugarcane varieties are Co 1148, Co 1158, Co 64, Co 997, Co 6304, Co 419, Co 740, Bo 17 and Co Pant 84211. Some varieties with special characters are given below:

1. **Early varieties**: Co 8336, Co 8337, Co 8338, Co 8339, Co 8340, Co 8341, (8-9 months) Co J64, and Co C 671.
2. **Resistant to abiotic stresses**

   Drought: Co 285, Co 740, Co 997, Co 1148 Frost : Co 1148 and N Co 310.

   Salinity: Co 453 and Co 62125.

   Lodging: Co 6304, Co 7117 and CoS 7918.

   Water logging: Co 1157, Co 975, Co 785, Bo 91. Bo 104, Bo 106, Bo 109, Bo 120, Cos 510, CoS 767 and CoS 8016.
3. **Resistant to biotic stresses**

   Top borer: COJ 67, Co 1158 snd Co 7224.

   Internode borer: COC 671, Co 975, Co 6806, Co 62175, *etc.*

   Red rot: Co 7627, COJ 64, COR 8001; CoLK 8001, COH 7803, Bo 108, *etc.*

Q15. Explain briefly about family, distribution and uses of Tobacco.

Ans. **(i) Family:** Tobacco belongs to the family *Solanaceae*. It is a narcotic plant which has commercial value. It is grown in tropical and subtropical climates the world over.

(ii) Global Distribution: The important tobacco growing countries are USA, China, India, Thailand, Pakistan, Myanmar, Sri Lanka, Indonesia, *etc.*

(iii) National Distribution: In India, the major tobacco growing states are Andhra Pradesh, Gujarat, Karnataka, Maharashtra, Tamil Nadu, Uttar Pradesh, Bihar and West Bengal.

(iv) Types: Cultivated tobacco is of two types, *viz. Nicotiana rustica* and *Nicotiana tabaccum*. Depending on use, tobacco is classified into various groups such as cigarette (flue cured) tobacco, Bidi tobacco, Hookah tobacco, Cigar and Cheroot type, chewing and snuff type, *etc.*

(v) Uses: Tobacco is used for chewing, smoking and medicinal uses. Smoking id done through cigarette, Bidi, Hookah, Cigar and Cheroot.

Q16. Explain briefly about cultivated species, origin and evolution of Tobacco.

Ans. There are two cultivated tobacco, *viz. Nicotiana rustica* and *Nicotiana tabaccum*. Both species are tetraploid [2n=4x=48]. *Nicotiana rustica* requires cooler climate whereas *N. tabaccum* can be grown throughout India. *Nicotiana tabaccum* is the predominant species which accounts for 90 per cent of the total tobacco area. The remaining 10 per cent is covered by *N. rustica*. There are more than 60 wild species of tobacco which are found in South and North America, Australia and South Pacific Islands. Wild species are important sources of resistance to biotic and abiotic stresses.

Origin and Evolution: Andes region is the original home of tobacco. There are two cultivated species of tobacco [*Nicotiana tabacum* and *N. rustica*]. Both these cultivated species have developed from a cross between two different wild diploid species as follows:

(*a*) *N. sylvestris* (2*n* = 24) x N. *tomentosa* (2*n* = 24) = *N. tabacum* = (2n= 48)

(*b*) *N. paniculata* x *N. undulata*. = *N. rustica* = (2n =48)

(i) Origin of Nicotiana tabacum

Parents	*Nicotiana sylvestris*	*x*	*Nicotian tomentosa*
Chromosome	[2n=24]	[2n=24]	
F1		[2n=24] Sterile	
Chromosome doubling in nature		[2n =48] *Nicotiana tabacum* [Tetraploid-Fertile]	

(ii) Origin of Nicotiana rustica

Parents	*Nicotiana paniculata*	*x*	*Nicotian undulata*
Chromosome	[2n=24]	[2n=24]	
F1		[2n=24] Sterile	
Chromosome doubling in nature		[2n =48] *Nicotiana rustica* [Tetraploid-Fertile]	

FIGURE 16.3. Probable Origin of Cultivated Tetraploid Species of Tobacco.

Clausen and Goodspeed [1925] synthesized a new hexaploid species of tobacco (*Nicotiana*) from a cross between *Nicotiana tabacum* (2*n* - 48) and *N. glutinosa* (2*n* = 24). The *F*1 was sterile with 2*n* = 36, which made fertile by doubling of chromosome through colchicine treatment The new species, is known as *N. digluta.*

Parents	*Nicotiana tabacum x*	*Nicotian sylvestris*
Chromosome	[2n=48]	[2n=24]
F1	[2n=36] Sterile	
Chromosome doubling	[2n =72] *Nicotiana digluta* [Hexaploid-Fertile]	

FIGURE 16-4. Probable Origin of Hexaploid Species of Tobacco.

Q17. Explain briefly about reproduction and pollination of Tobacco.

Ans. Tobacco is a self-pollinated and seed propagated crop. Main points are presented as follows:

(i) Flowers: Flowers are bisexual and are called as arrow. Flowering requires warm nights, humid conditions and high rain fall. These conditions are available at Coimbatore.

(ii) Pollination: Tobacco is a self-pollinated crop. Outcrossing is less than 5 per cent.

(iii) Apomixis: In tobacco apomixis is not reported.

(iv) Propagation: Tobacco is a seed propagated crop.

(v) Isolation Distance: An isolation distance of 5 m is required for the foundation of and certified seed.

Q18. What are Breeding Objectives in Tobacco?

Ans. In tobacco, major breeding objectives are higher yield, acceptable quality and resistance to diseases and insects as follows:

(i) Higher Leaf yield: Major yield components are number of leaf per plant, leaf size and weight of single leaf.

(ii) Acceptable quality. Quality characters include nicotine content, sugar content and aroma. High nicotine is preferred in Bidi, hookah and chewing tobacco and low in cigarette tobacco. High sugar content is preferred in cigarette tobacco. In USA, cigarette tobacco contains about 18 per cent sugar. Thin leaves are preferred for cigar and pipe smoking, thicker leaves for cigarettes and the thickest leaves for chewing.

(iii) Disease Resistance: Major diseases black shank, bacterial blight, Fusarium wilt, powdery mildew, anthracnose, mosaic leaf curl, *etc.*

(iv) Insect Resistance: The main insect is tobacco bud worm.

Q19. Explain briefly the breeding procedures used in Tobacco.

Ans. The main breeding methods used in tobacco are of two types, *viz.* general breeding methods and special breeding methods as follows:

(i) General Methods: General methods include, introduction, pure line selection, mass selection, backcross, pedigree selection and single seed descent method.

(ii) Special Methods: Such methods include mutation breeding, distant hybridization and biotechnology. Mutation breeding is used especially for creating variability for various economic characters. Distant hybridization is used for interspecific gene transfer. Biotchnology especially tissue culture technique is useful in developing haploids, and interspecific gene transfer adopting embryo rescue and protoplast fusion techniques.

Q20. What are breeding centers of Tobacco?

Ans. In India, tobacco improvement work is carried out by following organizations:

(i) Central Tobacco Research Institute, (CTRI) Rajahmundry, Andhra Pradesh,

(ii) All India Coordinated Tobacco Improvement Project (AICTIP), Anand (Gujarat) and

(iii State Agricultural Universities in tobacco growing states. New varieties are released through coordinated project after multi-locational testing for 3-5 years.

Q21. What are Practical Achievements of Tobacco breeding?

Ans. In India, several high yielding and good quality varieties of tobacco suitable for various uses have been released for different states through CTRI, AICTIP and SAUs. Some recently released varieties are listed below:

(i) Flue cured varieties: CTRI spl, Jayasri, Godavari spl, Swarna, Hema, Bhavya, *etc.*

(ii) Cheroot: Bhavani spl, Lanka spl, Viswanadh, Prabhat, *etc.*

(iii) Chewing: Thangam, Bhagyalakshmi, Maragandham, Sona, *etc.*

(iv) Cigar filters: Krishna.

Q22. Explain briefly about family, distribution and uses of Potato.

Ans. (i) Family: Potato belongs to the family *Solanaceae*. It is an important vegetable crop which is grown in temperate regions the world over.

(ii) Global Distribution: Europe and Asia together account for 85 per cent of world potato production. Main potato producing countries are USA, Russia, India, China, *etc.*

(iii) National Distribution: In India, the major potato growing states are Himacha Pradesh, Punjab, Uttar Pradesh and Bihar.

(v) **Uses**: Potato is used as important vegetable crop. Hence, it is called as king of vegetables. Several products are made from potato.

Q23. Explain briefly about cultivated species, origin and evolution of Potato.

Ans. There are four types of cultivated potato, *viz.* diploid [2n=24], triploid [3x =36], tetraploid [4x = 48] and pentaploid [5x =60]. The most important cultivated species is *Solanum tuberosum* which is grown through the World. *S. tuberosum* is cultivated mainly in high hills of South America. Diploid, triploid and pentaploid species are cultivated in Peru and Bolivia. Several wild species of potato are found in USA and Mexico.

Origin and Evolution: The Central Andean region of South America is the original home of potato. Cultivated potato [*Solanum tuberosum*] is a tetraploid species [2n=48] which is believed to have originated from the diploid species *S. stenotomum* which is also found growing in the Andes region of South Peru and Bolivia. Eralier it was believed that tetraploi cultivated species *S. tuberosum* has originated through chromosome doubling from wild diploid species *S. stenotomum* found in the Andes region of South Peru and Bolivia. According to recent view, it is probably a segmental allopolyploid derived from the cross between cultivated diploid species *S. stenotomum* and wild diploid species *S. sparsipillum* [Hawkes, 1956, 58].

(i) Earlier view

Solanum stenotomu → chromosome doubling → Solanum tuberosum

(ii) Recent View

Parents	*Solanum stenotomum*	x	*S. sparsipillum*
Chromosome	[2n=24]		[2n=24]
F1		[2n=24] sterile	
Chromosome doubling in nature		[4n =48] *Solanum tuberosum* fertile	

FIGURE 16-5. Probable Origin of Cultivated Tetraploid Potato.

Q24. Explain briefly about reproduction and pollination in Potato.

Ans. Potato is a self-pollinated and vegetatively propagated crop. Main points are presented as follows:

(i) Flowers: Flowers are bisexual.

(ii) Pollination: Potato is a self-pollinated crop.

(iii) Self-incompatibility: In potato self- incompatibility is reported.

(iv) Propagation: Potato propagates by tubers.

Q25. What are the Breeding Objectives in Potato?

Ans. In potato, major breeding objectives are higher yield, quality, earliness, wider adaptation, resistance to diseases, insects, herbicides, heat, drought and frost as follows:

(i) **Higher Tuber yield**: Major yield components are number of tubers per plant and tuber weight.

(ii) **Acceptable quality:** Quality characters include shape, size, , color, skin thickness, position of eyes and nutritive value. Tubers with external (cracks, knobs) and internal (hollow heart, necrosis and pigmentation) defects are discarded. High solid contents, good flavour and texture and low glycoalkaloid contents come under quality characters.

(iii) **Disease Resistance**: Major diseases are late blight, potato virus x and y, charcoal rot, potato wart and early blight.

(iv) **Insect Resistance**: The main insects are aphids, beetle and nematodes. is tobacco bud worm.

Q26. What are the Breeding procedures used in Potato?

Ans. Breeding methods which are used in potato are of two types, *viz.* general methods and special methods as follows:

(i) **General Methods**: Such methods include introduction, clonal selection, backcrossing and recurrent selection.

(ii) **Special Methods:** Such methods include interspecific hybridization and induced mutations. In potato, superior genotype can be selected at any stage and can be maintained easily by vegetative reproduction. There is no need for a selected individual to breed true. Asexual propagation permits fixation of heterosis. Recurrent selection is used for improvement of parental populations.

Biotechnology is expected to play an important role in future in potato improvement. Somaclonal variation will be useful in developing varieties resistant to biotic and abiotic stresses including herbicides. Biotechnology will also help in interspecific gene transfer.

Q27. What are breeding centers of Potato?

Ans. There are two types of potato breeding centers as follows:

(i) **International Centers:** International potato Centre (CIP), Peru serves as a world centre for genetic improvement of potato. The global collection of potato is also maintained at this centre. The main focus of this centre is on the improvement of potato for tropical and temperate climates of different countries. It has testing centres in different countries.

(ii) **National Centers:** In India, potato breeding work is carried out by following organizations:

(a) Central Potato Research Institute (CPRI), Shimla (Himachal Pradesh). Potato requires long day conditions, and monsoon climate for flowering. These conditions are available in Shimla. This is why potato Research Institute has been located in Shimla. CPRI has its regional stations in Ooty, Darjeeling and Mukteshwar (Nainital) in the hill regions, besides centres in plains at Jalandhar, Meerut, Patna, Bangalore, Gwalior. The seed multiplication work is carried out at Modipuram (Meerut).

(b) State Agricultural Universities.

The new varieties are released by All India Coordinated Potato Improvement Project (AICPIP), Shimla through multi-location testing for 3-5 years.

Q28. What are Practical Achievements of Potato breeding?

Ans. CPRI, Shimla has done remarkable work on potato improvement. Several high yielding varieties for various areas have been developed by this Institute. Some important varieties are Kufri Red, Kufri Sindhuri, Kufri Chamatkar, Kufri Kuber, Kufri Kundan, Kufri Alankar, Kufri Chandramukhi and Kufri Jyoti.

Chapter 17

Breeding Oilseed Crops

Q1. Explain briefly about family, distribution and uses of Rapeseed and mustard.

Ans. There are several oil yielding crops such as rapeseed and mustard, ground nut, soybean, sunflower, *etc.*

(i) **Family**: Rapeseed and mustard are important oilseed crops which belong to the family *Cruciferae*. They contribute more than 13 per cent to the global production of edible oil. Rapeseed and mustard collectively are known as oilseed brassicas. *Brassicas* can be grown in cool season subtropics, higher elevations and as winter crops in the milder region of temperate zone.

(ii) **Global Distribution**: Major rapeseed and mustard growing countries are China, Canada, the Indian subcontinent and northern Europe. China, India and Pakistan collectively contribute more than 90 per cent to the global production.

(iii) **National Distribution**: In India, major *Brassica* producing states are, Uttar Pradesh, Rajasthan, Punjab, Assam, Bihar and West Bengal.

(iv) **Uses**: The seeds of *brassica* contain 40 to 45 per cent oil and 38 to 41 per cent protein. In India, oil is used for human consumption and cake as livestock feed.

Q2. Explain briefly about cultivated species, origin and evolution of Rapeseed and mustard.

Ans. (i) **Cultivated Species:** There is one main cultivated species of rapeseed *i.e. Brassica campestris* (2n=20). This species has three ecotypes, *viz.* yellow sarson (*Brassica campestris* var. yellow sarson), brown sarson (*Brassica campestris* var. brown sarson) and toria (*Brassica campestris* var. *toria*). Yellow sarson and brown sarson are collectively called as turnip rape and Toria is known as Indian rape. The other cultivated species is *Brassica napus* (2n=38). In India, there are two main cultivated species of mustard, *viz.* Rai or Indian mustard (*Brassica juncea* 2n=4x=36) and Banarsi Rai or Black mustard (*Brassica nigra* 2n=16). There are 36 species in the genus Brassica. Most of the species are non oleiferous.

(ii) Origin and Evolution: It is bel *Brassica juncea* ieved that Rapeseed has (*Brassica campestris*) has originated in Himalayan region. Middle East is considered as primary cenre of origin of Indian Mustard. China, North East India and Caucasus are considered as secondary centres of origin. *Brassica juncea* is amphidiploid betwee *B. nigra* and *B. campestris*, *Brassica carinata* between *B. nigra* and *B. oleracea*, and *Brassica napus* between *B. oleracea* and *B. campestris*.

In *brassica*, there are three basic species, *viz.* black mustard [*Brassica nigra*] (*BB*, $2n = 16$), cabbage mustard [*B. oleracea* (*CC*, $2n = 18$)] and yellow sarson [*B. campestris* (*AA*, $2n = 20$)]. Three new tetraploid species have originated from a cross between two different diploid species as follows:

(*a*) *B. nigra* x *B. oleracea* = *B. carinata.*

(*b*) *B. compestris* x *B. oleracea* = *B. napus*

(*c*) *B. campestris* x *B. nigra* = B. *juncea.* [Indian mustard]

(i) Origin of Brassica carinata

Parents	*Brassica nigra*	x	*Brassica oleracea*
Genome	BB [2n=16]		CC[2n=18]
F_1		BC [2n=17] sterile	
Chromosome doubling in nature	BBCC [2n =34] *B. caranata*- fertile		

(ii) Origin of Brassica napus

Parents	*Brassica camestris*	x	*Brassica oleracea*
Genome	AA [2n=20]		CC [2n=18]
F_1	AC [2n=19] sterile		
Chromosome doubling in nature	AACC [2n =38] *B. napus*- fertile		

(iii) Origin of Brassica juncea

Parents	*Brassica camestris*	x	*Brassica nigra*
Genome	AA [2n=20]		BB [2n=16]
F_1	AB [2n=18] sterile		
Chromosome doubling in nature	AABB [2n =36] *B. juncea* -fertile		

FIGURE 17-1. Origin of Tetraploid Species of *Brassica* as per Nagharu U.

Q3. Explain briefly about reproduction and pollination of Rapeseed and mustard.

Ans. Rapeseed and mustard are seed propagated crops. Main points are presented as follows:

(i) **Pollination**: *Brassica napus* and *Brassica juncea* are self-pollinated species with natural outcrossing up to 30 per cent. Indian yellow sarson is self- compatible.

(ii) **Self-incompatibility**: Brown sarson and Toria are self-incompatible. *Brassica nigra* and *B. oleracea* are also self-incompatible. Self-incompatibility promotes cross pollination in these species which occurs by wind and honey bees.

(iii) **Propagation**: Rapeseed and mustard are seed propagated crops.

(iv) **Isolation Distance**: An isolation distance of 400 meters for the production of foundation seed and 200 meters for certified seed is required under Indian conditions.

Q4. What are Breeding Objectives of Rapeseed and mustard?

Ans. Main breeding objectives in rapeseed and mustard are higher yield, early maturity, shattering resistance, ideal plant type, better quality and resistance to biotic (diseases and insects) and abiotic (drought and temperature) stresses.

(i) **Higher grain yield**: Major yield components include number of primary and secondary branches, number of siliquae per plant, seed per siliqua and seed size.

(ii) **Quality Traits**: Quality includes high seed oil content and low erucic acid and glucosinolate in the oil.

(iii) **Disease Resistance**: Major diseases include alternaria blight, white rust, downey mildew, *etc.*

(iv) **Insect Resistance**: Aphid is the major insect pest of rapeseed and mustard.

Q5. Explain briefly the breeding procedures used in Rapeseed and mustard.

Ans. There are two types of rapeseed and mustard species, *viz.* self-compatible and self-incompatible (cross pollinated).

(i) **Self-Pollinated Species:** In self-pollinated species, pure line selection, back crossing, pedigree method, single seed descent method and bulk breeding methods are commonly used.

(ii) **Cross-Pollinated Species**: In cross-pollinated species, mass selection, progeny selection, recurrent selection and synthetic breeding are generally used.

(iii) Distant hybridization is used for interspecific or intergeneric gene transfer.

(iv) Natural polyploidy has resulted in the development of three amphidiploid species (*B. napus, B. juncea* and B. carinata). Mutation breeding is used for creating variability for various economic characters.

(v) Somaclonal variation recorded for earliness, dwarfness, high yield and resistance to diseases and drought (Chopra *et al.*, 1989. Haploids and transgenic plants have been developed through biotechnology.

Q6. What are breeding centers of Rapeseed and mustard?

Ans. In India, Rapeseed and mustard breeding work is carried out by National Research Centre for Mustard, Bharatpur (Rajasthan) and State Agricultural Universities of Rapeseed and Mustard growing states. New varieties are released after multi-location testing in coordinated trials for 3-5 years. The coordinated project was located at HAU, Hisar. Now this has been shifted to Bharatpur. Several high yielding varieties of rapeseed and mustard have been released in India for different states by above research centres.

Practical Achievements: Important National varieties of mustard are Kranti, RLM 198, Krishna and RLM 514. Pant Toria 303 is the national variety of rapeseed.

Q7. Explain briefly about family, distribution and uses of sunflower (*Helianthus annuus* L.)

Ans. **(i) Family**: Sunflower belongs to the family *Compositae*. It is an important oilseed crop of global importance. It contains 46 to 52 per cent oil.

(ii) Global Distribution: Important sunflower producing countries are USSR, USA, Argentina, China, Australia, France, Spain, Romania, Hungary, Yugoslavia, South Africa and Canada.

(iii) National Distribution: In India, it was introduced in 1969 from USSR. Now, it is cultivated in Karnataka, Maharashtra, Tamil Nadu and Andhra Pradesh. It is spreading in Punjab and Haryana states also.

(iv) Uses: Sunflower oil is used for human consumption and cake as cattle feed.

Q8. Explain briefly about cultivated species, origin and evolution of Sunflower.

Ans. There are two species of sunflower, *i.e. Helianthus annuus* ($2n = 34$) and *H. tuberosus* which are cultivated as food crops. *H. annuus* is grown as oilseed crop. There are 65 other species, out of which 8 are grown as ornamentals and rest are wild. The cultivated species is of two types, *viz.* oilseed type and non- oilseed type. The non-oilseed types are rich in protein and are grown in USA and some other countries. Non- oilseed types are used whole, dehulled, roasted or raw alone or added to other food.

Origin: Archaeological evidences suggest that Central America is the center of origin of sunflower. Nothing is known about the progenitor of sunflower so far.

Q9. Explain briefly about reproduction and pollination in Sunflower.

Ans. Sunflower is seed propagated and cross pollinated species. Natural cross pollination occurs mainly by insects. Thus sunflower is a highly heterozygous and heterogeneous crop. An isolation distance of 400 m and 200 m is essential for the production of foundation and certified seeds.

Q10. What are Breeding Objectives in Sunflower?

Ans. Main breeding objectives are higher yield, better quality, wider adaptation, and resistance to biotic (diseases and insects) and abiotic (drought and temperature) stresses.

(i) **Higher Yield**: Major yield components include head size, seeds per head and seed size.

(ii) **Quality Traits**: Quality refers to oil and protein content and their composition.

(iii) **Disease Resistance**: Major diseases include leaf spots, root rot, stem rot, head rot, rust and powdery mildew.

(iv) **Insect Resistance**: *Heliothis*, grasshoppers, jassids and leaf eating cater pillars are major insects.

Q11. Explain briefly the breeding procedures used in Sunflower.

Ans. Main breeding methods are mass selection, recurrent selection, heterosis breeding, synthetic breeding, *etc.* Haploids and transgenic plants have been developed through biotechnology.

Q12. What are breeding centers of Sunflower?

Ans. New varieties are developed by State Agricultural Universities and Private seed companies after multi-locational testing in Co-ordinated trials. All India Coordinated Sunflower Improvement Project is located at the University of Agricultural Sciences, Bangalore.

Practical Achievements: New varieties are released through Coordinated Project after multi-location testing for 3-5 years. Coordinated Project is located at Indore. Important sunflower varieties and hybrids released in India are listed below:

1. **Varieties:** EC 68414, EC 68415, Modern, Co 1, Co 2, Surya, SS 56, MSFS 1, MSFS 8 and MSFS 17.
2. **Hybrids:** BSH 1, APSH 11, LDMRSH 1, LDMRSH 3. All these Hybrids are CMS based.

Q13. Explain briefly about family, distribution and uses of Soybean.

Ans. (i) **Family**: Soybean belongs to the family *Leguminosea*. It is an important protein cum oil yielding crop of global importance.

(ii) **Global Distribution**: Important soybean producing countries are USA, Brazil, China, Argentina, Canada and Japan. USA alone contributes over 60 per cent to the global production of soybean.

(iii) **National Distribution**: In India, major soybean producing states are M. P. ,U. P., and Maharashtra.

(iv) **Uses**: In India, both soybean seed and oil are used for human consumption. Soybean contributes 60 per cent to vegetable protein and 30 per cent to edible oil on global basis.

Q14. Explain briefly about cultivated species, origin and evolution of Soybean.

Ans. Glycine max [2n=40] is the only cultivated species of soybean. This is an annual diploid species. There are 8 wild species of soybean out of which 7 are perennial and one (*Glycine soja*) annual diploid. Wild species are important sources of disease and drought resistance. Soybean is cultivated in tropical and sub-tropical climate.

Origin and Evolution: It is generally accepted that soybean originated in China from where it spread to East and South East Asia, USA and other countries. The wild annual diploid species *Glycine soja* is considered as the progenitor of cultivated soybean. It is also believed that annual form originated from *Glycine tabacina* or *G. tomentella* because these perennial species are also found growing in area of *G. soja.*

Q15. Explain briefly about reproduction and pollination in Soybean.

Ans. Soybean is self-pollinated and seed propagated crop. Main points are presented as follows:

(i) **Pollination**: Self-pollination. Flower are bisexual.

(ii) **Propagation**: Soybean is a seed propagated crop.

(iii) **Isolation Distance**: An isolation distance of 3 meters is safe for the production of foundation and certified seed.

Q16. What are Breeding Objectives of Soybean?

Ans. In soybean, main breeding objectives are higher yield, better quality, wider adaptation, and resistance to biotic (diseases and insects) and abiotic (drought and temperature) stresses.

(i) **Higher Yield**: Major yield components are pods per plant, seeds per pod and seed size.

(ii) **Quality Traits**: Quality includes oil and protein content and absence of anti-nutritional factors.

(iii) **Disease Resistance**: Major diseases include pod blight, yellow mosaic, bacterial blight and alternaria leaf spot.

(iv) **Insect Resistance**: *Bihar hairy caterpillar and girdle beetle are the main insects.*

Q17. Explain briefly the breeding procedures used in Soybean.

Ans. Breeding procedures which are used for soybean improvement are of two types, *viz.* general methods and special methods as follows:

(i) **General Methods**: These methods include introduction, pure line selection, pedigree, back cross and multiline breeding, bulk and single seed descent methods.

(ii) **Special methods**: Such methods include: mutation breeding, distant hybridization and transgenic breeding.

(iii) Somaclonal variation may be helpful in developing resistant genotypes to biotic and abiotic stresses and improving several other traits.

Q18. What are breeding centers of Soybean?

Ans. Soybean improvement work is carried out by various International and National crop research centers as follows:

(i) **International Centers:** Following Internal crop research centers are involved in soybean improvement.

(a) AVRDC: Asian Vegetable Research and Development Center [AVRDC], Taiwan. AVRDC deals with improvement of tomato, Chinese cabbage, sweet potato, mung bean and soybean for Asian region.

(b) IITA: International Institute of Tropical Agriculture, Nigeria. IITA deals with improvement of cowpea, soybean, maize, rice, root tubers and plantation crops.

(ii) **National Centers:** In India, soybean improvement work is carried out by following organizations:

(a) Directorate of Soybean Research, Indore, Madhya Pradesh. The new varieties are released through coordinated project after multi-location testing for 3-5 years.

(b) State Agricultural Universities located in soybean growing regions.

Q19. Describe briefly practical achievements of Soybean breeding.

Ans. In India, several improved varieties of soybean have been developed for different states. Eight varieties (Bragg, Clark 63, Davis, Hardee, Improved Pelicon, Jupiter, Lee and Monetta) have been released through direct introduction. Some high yielding varieties have been developed by Pantnagar, Jabalpur, Ludhiana and other Agricultural universities and also by IARI, New Delhi. These varieties include PK 262, PK 308, PK 327, PK 416, PK 472, Shilajeet and Pant Soybean 564 (Pantnagar); Durga, Gaurav, JS 75-46, JS 80-21, (*Jabalpur); SL 4 and SL 96 (Ludhiana); Pusa 16 Pusa 20 and Pusa 24 (IARI, New Delhi); VL Soya 2 (Almora), Shivalik (Palampur); and MACS 13 and MACS 58 (Poona).

Q20. Explain briefly about family, distribution and uses of Groundnut.

Ans. **(i) Family:** Ground nut belongs to the family *Leguminosea*. It is an important oil yielding crop of global importance.

(ii) **Global Distribution**: Peanut is a tropical crop, but its cultivation extends to both tropical and sub- tropical regions of the world.

Principal peanut producing countries are India, China, USA, Africa [Senegal and Nigeria] and South and South East Asia [Myanmar, Malayasia Pakistan, and Sri Lanka].

(iii) National Distribution: In India, major peanut producing states are Gujarat, Andhra Pradesh, Karnataks, Tamil Nadu, Maharashtra, Madhya Pradesh, Rajasthan, Uttar Pradesh, Odisha and Punjab.

(iv) Uses: Kernels are rich in oil, protein, vitamins and minerals and are used raw or roasted for human consumption. Leaves and cake are used as cattle feed shells as fuel.

Q21. Explain briefly about cultivated species, origin and evolution of Groundnut.

Ans. *Arachis hypogaea* [2n=40] is the only cultivated species of peanut. This is an annual tetraploid species. This species has two sub-species, *viz. Arachis hypogaea* ssp *hypogea* and *A. hypogaea* ssp *fastigiata*. The former sub-species has no floral axis on the main stem, while the latter has. Each sub-species has two types as follows:

(i) Sub-species hypogaea: It has two types, *viz.* hypogaea and hirsute. The variety hypogaea has lesser hair and shorter branches than hirsute. The variety hypogaea is of two types, *viz.* small poded with spreading (runner) growth habit called runner, and large poded with erecthabit called Virginia. The hirsute is not much cultivated.

(ii) Sub-species fastigiata: It is of two types, *viz.* fastigiata or vanecia and vulgaris or Spanish. The Valencia is large poded and Spanish has small pods.

There are about 15 wild species of peanut which are used in interspecific gene transfer for resistance to biotic and abiotic stresses.

Origin and Evolution: It is generally accepted that peanut originated Brazil from where it spread to other countries. The tetraploid wild species Arachis monticola [2n= 4x=40] is considered to be the progenitor of cultivated peanut. The wild species Arachic monticola has several similarities with cultivated species. Moreover, it can be easily crossed with cultivated peanut.

Q22. Explain briefly about reproduction and pollination in groundnut.

Ans. Cultivated peanut is self-pollinated and seed propagated crop. Main points are presented as follows:

(i) Pollination: Peanut is a self-pollinated species.The percentage of natural cross pollination is practically negligible.

(ii) Propagation: Peanut is a seed propagated crop. Some wild species propagate vegetativelt usually by rhizomes.

(iii) Isolation Distance: An isolation distance of 3 meters is safe for the production of foundation and certified seed.

Q23. What are the Breeding Objectives in groundnut?

Ans. In peanut, main breeding objectives are higher yield, better quality, wider adaptation, and resistance to biotic (diseases and insects) and abiotic (drought and temperature) stresses.

(i) **Higher grain Yield:** Major yield components are pods per plant, kernels per pod, kernel size and shelling per cent.

(ii) **Quality Traits:** Quality includes seed oil and protein content and composition flavor and milling quality.

(iii) **Disease Resistance**: Major diseases include leaf spot, dry root rot, Fusarium wilt, mosaic virus and rosette.

(iv) **Insect Resistance:** Important insects include, red caterpillar, Bihar hairy caterpillar, leaf miner, pod borer and pod sucking bugs.

Q24. Explain briefly the breeding procedures used in Groundnut.

Ans. Breeding procedures which are used for peanut improvement are of three types, *viz.* general methods, special methods and population improvement procedures as follows:

(i) **General Methods**: These methods include introduction, pure line selection, pedigree, back cross and multiline breeding, bulk and single seed descent methods.

(ii) **Special methods**: Such methods include: mutation breeding, distant hybridization and transgenic breeding.

(iii) **Population Improvement**: Three breeding procedure, *viz.* biparental mating, disruptive mating and diallel selective mating are useful in breaking linkage blocks and creating vast genetic variability in a population.

(iv) **Biotechnology**: Biotechnological tools such as embryo rescue and tissue culture has made it possible to utilize genes from wild species.

Q25. What are breeding centers of Groundnut?

Ans. Peanut improvement work is carried out by various International and National crop Research centers as follows:

(i) **International Centers:** International Crop Research Institute for Semi-arid Tropics [ICRISAT], Hyderabad is involved in peanut improvement.

(ii) **National Centers:** In India, soybean improvement work is carried out by following organizations:

(a) Directorate of Groundnut Research, Junagarh, Gujarat.

(b) State Agricultural Universities located in peanut growing regions.

The new varieties are released through coordinated project after multi-location testing for 3-5 years.

Q26. Describe briefly practical achievements of Groundnut breeding.

Ans. In India, more than 60 improved varieties of peanut have been released for cultivation in different states. List of important peanut varieties released in Spanish, Valencia and Virginia groups in India is presented as follows:

1. **Spanish Type**: MH 1, SG 84,GG 2, J 11, Jyoti, TG 17, Javan, Kisan, ICGS 11, Co 1, Co 2, TMV 5, TMV 7, TMV 9, TMV 12, *etc.*
2. **Valencia Type**: MH 2, Kopergaon 3, TMV 3, *etc.*
3. **Virginia Type**: T 28, T 64, M 145, M197, TMV 6, TMV 8, TMV 10, Kausal, Kadri 2 and Kadri 3 are semi-spreading types. Spreading types include M 13, M37, M335, GG 11, TMV 1, TMV 3, TMV 4, S 230, Kadri 71-1, *etc.*

Chapter 18

Breeding Pulse Crops

Q1. Give a brief classification of pulse crops.

Ans. Those Leguminous species whose seeds are used for human consumption are referred to as pulses or grain legumes (Roberts, 1970). Important pulse crops include chickpea, field pea, pigeon pea, black gram, green gram, lentil, horse gram, moth bean, field bean, *khesary*, cowpea, cluster bean, *etc*. Pulse crops belong to the family Fabaceae (old Leguminosae).

Types: Pulse crops are classified in various ways as presented in the Table 18.1.

TABLE 18-1. Classification of Pulse Crops

Sl.No.	*Basis of Classification*	*Categories*	*Examples of pulse crops*
1	Utilization	Major or Primary Species	Chickpea, pea, lentil, pigeon pea, green gram, black gram and cowpea.
		Minor or Secondary Species	Horse gram, moth bean, field bean, *khesary*, and cluster bean.
2	Temperature tolerance	Cool season Species	Chickpea, pea and lentil.
		Warm season Species	Pigeon pea, green gram, black gram and cowpea.
3	Growing season	*Kharif* season	Chickpea, pea and lentil.
		Rabi season	Pigeon pea, green gram, black gram and cowpea.

(i) Cool Season Pulses: Such species include chickpea, field pea, lentil and *khesary* are cool season pulses of subtropical and mild temperate regions.

(ii) Warm season Pulses: Such species include Pigeon pea, green gram, black gram and cowpea. Rest are warm season pulses which are grown in tropical and subtropical climates.

Based on utilization, pulse crops are again of two types as follows:

(i) Primary Pulses: This group includes chickpea, pea, lentil (cool season), pigeon pea, green gram, black gram and cowpea (warm season). These are also known as major pulse crops.

(ii) Secondary Pulses: This group includes horse gram, moth bean, field bean, *khesary*, and cluster bean. These are also known as minor pulses.

Q2. Give in brief Global and National distribution of primary pulse crops.

Ans. A brief description and breeding of primary pulses is presented here. Major producing countries and Indian states for primary pulses are given in Table 18.2.

TABLE 18-2. Distribution of Primary Pulse Crops

Sl.No.	*Pulse Crop*	*Cultivated Species*	*Major Growing Countries*	*Major Growing Indian States*
1	Chick pea	*Cicer arietinum*	India, Pakistan, Turkey, Ethiopia and Burma	M.P. , Rajasthan, U. P. and Haryana
2	Field pea	*Pisum sativum*	China, India, Pakistan, USA, and Canada,	U. P., M. P. Rajasthan and Bihar
3	Lentil	*Lens culinaris*	India, Pakistan, Egypt and Greece	U. P., M. P. Bihar and West Bengal.
4	Pigeon pea	*Cajanus cajan*	India, Uganda, Kenya, West Indies, Burma *etc.*	Maharashtra, U. P. , M. P. Karnataka, Gujarat and Bihar.
5	Black gram	*Vigna mungo*	India, Pakistan, Bangladesh, Sri Lanka, Philippines, Taiwan, Thailand, Nepal *etc.*	M. P. Maharashtra, , Tamil Nadu, U. P. , A. P. *etc.*
6	Green gram	*Vigna radiata*	India, Pakistan, Bangladesh, Sri Lanka, Philippines, Taiwan, Thailand, Nepal *etc.*	Odisha, A. P., M. P., Maharashtra and Rajasthan.
7	Cow pea	*Vigna anguiculata*	Nigeria, Niger, Ghana, Kenya, Uganda, Tanzania, India, Sri Lanka, Burma, *etc.*	Rajasthan, M. P., U. P., Maharashtra and Gujarat.

Above pulses contain 18 to 27 per cent protein. The lowest protein is found in chickpea and field pea (18-20 per cent) and the highest in green gram and cowpea (24-27 per cent). Dry seeds are used as Dal, green pods as vegetables and leaves and stalks as cattle feed. Moreover, pulses fix 30 to 50 kg nitrogen per hectare in the soil per season.

Q3. Explain briefly place of origin and progenitor of primary pulse crops.

Ans. The cultivated species, center of origin and probable progenitors of primary pulse crops are given in Table 18-3.

TABLE 18-3. Place of Origin and Probable Progenitor of Primary Pulse Crops

Sl.No.	Name of Crop	Cultivated Species	Place of Origin	Progenitor
1	Chick pea	*Cicer arietinum*	South Turkey and Syria	*Cicer reticulatum*
2	Field pea	*Pisum sativum*	Mediterranea, Ethiopia and Near East	Not known
3	Lentil	*Lens culinaris*	Near East	*Lens orientales*
4	Pigeon pea	*Cajanus cajan*	Africa and India	*Cajanus cajanifolius*
5	Black gram	*Vigna mungo*	Indian sub-continent	*Vigna radiate* var *sublobatis*
6	Green gram	*Vigna radiata*	India	*Vigna radiate* var *sublobatis*
7	Cow pea	*Vigna anguiculata*	Africa (PC), India, China (SC)	Not known

Q4. Explain briefly the mode of reproduction and pollination in primary pulse crops.

Ans. All primary pulses are seed propagated and self-pollinated. Other points are as follows:

(i) Flowers: Flowers are bisexual and amenable for crossing.

(ii) Pollination: All pulses are self-pollinated. Pigeon pea is often cross pollinated species in which natural cross pollination up to 7 per cent has been reported under Indian conditions.

(iii) Propagation: All pulses are seed propagated.

(iv) Isolation Distance: Isolation distance of 20 m. in chickpea, field pea, lentil, black gram and green gram; 50 m. in cowpea; and 200 m. in pigeon pea is required for the production of foundation and certified seeds.

Q5. What are the breeding objectives in pulse crops?

Ans. In grain legumes, major breeding objectives are, high yield, wider adaptation, shattering resistance, better quality, early maturity, resistance to biotic and abiotic stresses as follows:

(i) Higher grain yield: Major yield components are, number of pods per plant, pod length, seeds per pod and seed size.

(ii) Better quality: Main quality characters are seed color, seed size, protein content, methionine content, and low amount of anti-nutritional substance.

(iii) Disease Resistance: *Fusarium* wilt and rust are major diseases of chickpea, pea, and lentil. Yellow mosaic, bacterial blight, leaf spot, powdery mildew and root rot are common in pigeon pea, black gram, green gram and cowpea. In Pigeon pea, *Fusarium* wilt and sterility mosaic are also major diseases.

(iv) Insect Resistance: Pod borer, cutworm and aphids in rabi pulses; and pod borer, hairy cater pillar and jassids in Kharif pulses (except pigeon pea) are the major insect pests.

Q6. What are the breeding procedures used in pulse crops?

Ans. Breeding procedures which are used for improvement of pulse crops are of three types, *viz.* general methods, special methods and population improvement procedures as follows:

(i) **General Methods**: These methods include introduction, pure line selection, pedigree, back cross and multiline breeding, bulk and single seed descent methods.

(ii) **Special methods:** Such methods include: mutation breeding, distant hybridization, heterosis breeding and transgenic breeding. Singh (1991) has reviewed the mutation breeding work on pulse crops. Mutation breeding has helped in developing improved varieties in various pulse crops such as chickpea (Pusa 408, Pusa 413, Pusa 417 and Kiran), Lentil (Arun), field pea (Hans), Pigeon pea (Co 5, Co 3, TAT 5 and Trombay), black gram (Co 4), green gram (Co 4, ML 26-1-3, Pant Mung 2 and TAP 7) and cowpea (V 16, V 3, V 38 and V 240). In Pigeon pea, heterosis breeding is also used. Biotechnology will supplement to conventional methods in pulse crops.

(iii) **Population Improvement Procedures**: Such procedures include recurrent selection and diallel selective mating systems. These are used for specific purposes.

Q7. What are the breeding centers of pulse crops?

Ans. There are two types of pulse breeding centers as follows:

(i) **International Centers:** Following International Research Centers are engaged in improvement of pulse crops:

(a) **ICRISAT**: International Crop Research Institute for Semiarid Tropics, Hyderabad, India is actively engaged in improvement of pigeon pea and chickpea (both desi and Kabuli).

(b) **ICARDA:** International Centre for Agricultural Research in Dryland Areas (ICARDA) works on improvement of lentil and Kabuli chickpea.

(c) **AVRDC:** Asian Vegetable Research and Development Centre, Taiwan deals with breeding of mung bean and maintains global gene pool of this pulse.

(d) **CIAT**: International Centre for Tropical Agriculture (CIAT), Colombia deals with breeding of French bean.

(e) **IITA**: International Institute of Tropical Agriculture (IITA), Nigeria deals with improvement of Cowpea.

(ii) **National Centers:** In India, pulse breeding work is carried out by following organizations:

(a) Indian Institute of Pulses Research Kalyanpur, Kanpur The new varieties are released through coordinated project after multi-location testing for 3-5 years.

(b) State Agricultural Universities located in wheat growing regions.

Improvement of grain type cowpea has been included in the ICAR Coordinated Project on underutilized unexploited crops. Vegetable types are dealt with by Indian Institute of Vegetable Research, Varanasi and work on fodder legumes is carried out at Indian Grassland and Fodder Research Institute, Jhansi.

Q8. What are the Practical Achievements of pulse crops breeding in India?

Ans. In India, several high yielding varieties of grain legumes have been released for different states. ICRISAT has developed the first pigeon pea hybrid (ICPH 8) in the world for commercial cultivation in India. This hybrid has been developed using genetic male sterility. It is early maturing (140 days) and performs well under drought and high moisture conditions. This hybrid has been recommended for Maharashtra and Gujarat states. A list of important varieties released in various pulse crops in India is given below:

1. **Pigeon pea:** T 21, T 7, T 17, Prabhat, UPAS 120, Manak, Vishaka, ICPL 87, ICPL 151, LRC 30, LRC 36, Pusa 84, Pusa 85, *etc.*
2. **Chick pea:** T 3, T 4, T 5, Pant 114, K 850, H 208, H 355, Pusa 209, Pusa 240, Pusa 244, Pusa 256, Pusa 413, Pusa 417, Pusa 212, Avrodhi, *etc.*
3. **Lentil:** T 8, T 36, Pant 406, Pant L 639, L 9-12, Lens 4076, Malika, Pusa 4, Pusa 6, *etc.*
4. **Pea:** Type 163, Pant P 5, Rachna, DMR 11, Pant P 5, Hans, VL 1, B 12, *etc.*
5. **Black gram:** T 9, T 27, T 65, PS 1, Pant 419, 430, Khargone 3, Co 1, Co 2, Co 3, Co 4, Co 5, *etc.*
6. **Green gram:** P 37, P 310, PS 16, Pusa 105, Pusa Baisakhi, T 44, T 51, K 851, Plant Moong 2, Plant Moong 3, *etc.*
7. **Cowpea:** Pusa 152, Pusa Sawani, V 16, V 240, Co 1, Co 2, Co 3, Co 4, K 11, K 14, T 2, JC 5, JC 10, *etc.*

The yield of pulse crops is lower than cereals. Main reasons of low yield of pulse crops are: (i) long continued natural selection, (ii) cultivation on poor lands, (iii) inadequate application of fertilizers, (iv) inadequate plant protection measures, and (v) rain-fed cultivation.

Section IV

Miscellaneous Topics

Chapter 19

Ideotype Breeding

Q1. What is ideotype breeding?

Ans. In broad sense an ideotype is a biological model which is expected to perform or behave in a predictable manner within a defined environment. More specifically, crop ideotype is a plant model which is expected to yield greater quantity of grains, fibre, oil or other useful product when developed as a cultivar. The term ideotype was first proposed by Donald in 1968 working on wheat.

Q2. What are main features of plant ideotypes?

Ans. The main points about plant ideotype are given below:

(i) Crop ideotype refers to model plants or ideal plant type for a specific environment.

(ii) Ideotype differs from idiotype. The former refers to a combination of various plant characters which enhance the yield of economic produce, whereas the latter refers to the morphological features of the chromosomes of a particular plant species.

(iii) Donald included only morphological characters to define an ideotype of wheat. Subsequently, physiological and biochemical traits were also included for broadening the concept of crop ideotype.

(iv) Ideal plants or model plants are expected to give higher yield than old cultivars in a defined environment.

(v) Ideotype is a moving goal which changes according to climatic situations, type of cultivation, national policy, market requirement *etc.* In other words, ideotypes have to be redesigned depending upon above factors. Thus, development of crop ideotypes is a continuous process.

(vi) Ideal plant type or model plant type also varies from species to species (see later). Moreover, this is a difficult and slow method of cultivar development because various morphological, physiological and biochemical characters have to be combined in single genotype from different sources.

Q3. What is ideotype breeding?

Ans. Ideotype breeding or plant type breeding can be defined as a method of crop improvement which is used to enhance genetic yield potential through genetic manipulation of individual plant character. Each character plays a definite role in the enhancement of yield. In other words, plant characters are chosen in such a way that each character contributes towards increased economic yield. Main features of ideotype breeding are briefly discussed below:

1. **Emphasis on Individual Trait:** In ideotype breeding, emphasis is given on individual morphological and physiological trait which enhances the yield. The value of each character is specified before initiating the breeding work.
2. **Includes Yield Enhancing Traits:** Various plant characters to be included in the ideotype are identified through correlation analysis. Only those characters which exhibit positive association with yield are included in the model.
3. **Exploits Physiological Variation:** Genetic differences exist for various physiological characters such as photosynthetic efficiency, photo respiration, nutrient uptake, *etc.* Ideotype breeding makes use of genetically controlled physiological variation in increasing crop yields, besides various agronomic traits.
4. **Slow Progress :** Ideotype breeding is a slow method of cultivar development, because incorporation of various desirable characters from different sources into a single genotype takes long time. Moreover, sometimes undesirable linkage affects the progress adversely.
5. **Selection:** In ideotype breeding selection is focussed on individual plant character which enhances the yield.
6. **Designing of Model:** In ideotype breeding, the phenotype of new variety to be developed is specified in terms of morphological and physiological traits in advance.
7. **Interdisciplinary Approach:** Ideotype breeding is in true sense an interdisciplinary approach, it involves scientist from the disciplines of genetics, breeding, physiology, pathology, entomology *etc.*
8. **A Continuous Process:** Ideotype breeding is a continuous process, because new ideotypes have to be developed to meet changing and increasing demands. Thus development of ideotype is a moving target. Ideotype breeding differs from traditional breeding in the sense that

values for individual traits are specified in case of ideotype breeding, whereas such values are not fixed in case of traditional breeding. In other words, first a conceptional plant model is fixed and then efforts are made to achieve such model. In traditional breeding, such models are not developed before initiation of breeding program. There are several differences between traditional breeding and ideotype breeding.

Q4. Explain briefly features of crop ideotypes.

Ans. The crop ideotype consists of several morphological and physiological traits which contribute for enhanced yield or higher yield than currently prevalent crop cultivars. The morphological and physiological features of crop ideotype differ from crop to crop and sometimes within the crop also depending upon whether the ideotype is required for irrigated cultivation or rainfed cultivation. Ideal plant types or model plants have been discussed in several crops like wheat, rice maize, barley, cotton and beans. The important features of ideotype for some crops such as wheat, barley, rice, maize, cotton, sorghum and pearl millet are briefly described below:

Q5. Describe briefly main features of wheat ideotype.

Ans. The term ideotype was coined by Donald in 1968 working on wheat. He proposed ideotype of wheat with following main features:

1. **A short strong stem.** It imparts lodging resistance and reduces the losses due to lodging.
2. **Erect leaves.** Such leaves provide better arrangement for proper light distribution resulting in high photosynthesis or CO2 fixation. The conceptual theoretical model is prepared before initiation of breeding work. Selection is focussed on individual plant characters. It includes various morphological, physiological and biochemical plant characters. Value of each trait is defined in advance. This is a difficult and slow method of cultivar development. Phenotype of new variety to be developed is specified in advance.

(i) The main objective is defined before initiating the breeding work.

(ii) Selection is focused on yield and some other characters.

(iii) It usually includes various morphological and economic characters.

(iv) Value of each character is not fixed in advance.

(v) This is a simple and rapid method of cultivar development.

(vi) The phenotype of a new variety is not specified in advance.

3. **Few small leaves.** Leaves are the important sites of photosynthesis, respiration and transpiration. Few and small leaves reduce water loss due to transpiration.
4. **Larger ear.** It will produce more grains per ear.

5. **An erect ear.** It will get light from all sides resulting in proper grain development.
6. **Presence of awns.** Awns contribute towards photosynthesis.
7. A single culm.

Thus, Donald included only morphological traits in the ideotype. However, all the traits were based on physiological considerations. Finlay (1968) doubted the utility of single culm in wheat ideotype. He considered tillering as an important feature of wheat ideotype. Some workers suggested that a wheat plant with moderately short but broad flag leaf, long flag leaf sheath, short ear extrusion with long ear, and moderately high tillering capacity should give high yield per plant (Hsu and Watson, 1971). Asana proposed wheat ideotype for rainfed cultivation. Recent workers included both morphological and physiological characters in wheat ideotype.

Q6. Describe briefly main features of rice ideotype.

Ans. The concept of plant type was introduced in rice breeding by Jennings in 1964, though the term ideotype was coined by Donald in 1968. He suggested that in rice an ideal or model plant type consists of (1) semi dwarf stature, (2) high tillering capacity, and (3) short, erect, thick and highly angled leaves (Jennings, 1964, Beachell and Jennings, 1965). Jennings also included morphological traits in his model. Now emphasis is also given on physiological traits in the development of rice ideotype.

Q7. Describe briefly main features of maize ideotype.

Ans. In 1975, Mock and Pearce proposed ideal plant type of maize. In maize, higher yields were obtained from the plants consisting of (1) low tillers, (2) large cobs, and (3) angled leaves for good light interception. Planting of such types at closer spacings resulted in higher yields.

Q8. Describe briefly main features of barley ideotype.

Ans. Rasmusson (1987) reviewed the work on ideotype breeding and also suggested ideal plant type of six rowed barley. He proposed that in barley, higher yield can be obtained from a combination of (1) short stature, (2) long awns, (3) high harvest index, and (4) high biomass. Kernel weight and kernel number were found rewarding in increasing yield.

Q9. Explain briefly main features of cotton ideotype.

Ans. In cotton, genotypes with zero branch, short stature, compact plant, small leaves and fewer sympodia were considered to enhance yield levels. Singh *et al.* (1974) proposed an ideal plant type of uplant cotton (*Gossypium hirsutum*) and tree cotton (*G. arboreum*) for irrigated conditions of North Indian cotton growing belt. The proposed ideotype includes (1) short stature (90-120 cm), (2) compact and sympodial plant habit making pyramidal shape, (3) determinate in fruiting habit with unimodal distribution of bolling, (4) short duration (150-165 days), (5) responsive to high fertilizer dose, (6) high degree of inter plant competitive ability, (7) high degree of resistance to insect pests and diseases, and (8) high physiological efficiency. Singh and Narayanan (1993) proposed an ideotype of above two species for rainfed conditions. The main features of proposed ideotype include, earliness (150-165 days), fewer small and thick leaves, compact and short stature, indeterminate habit, sparse hairiness, medium to big boll size, synchronous bolling, high response to nutrients, and resistance to insects and diseases.

Q10. Describe briefly main features of Sorghum and pearl millet ideotype.

Ans. Improvement in plant type has been achieved in *Sorghum* and pearl millet through the use of dwarfing genes. In these crops dwarf *F*1 hybrids have been developed which have made combine harvesting possible.

Genetic improvement has been achieved through modification of plant type in several crop species. New ideotypes have been proposed for majority of crop plants. Swaminathan (1972) has listed several desirable attributes of crop ideotypes with special reference to multiple cropping in the tropics and sub tropics. These features include, (1) superior population performance, (2) high productivity per day, (3) high photosynthetic ability, (4) low photo respiration, (5) Photo and thermoinsensitivity, (6) high response to nutrients, (7) high productivity per unit of water, (8) multiple resistance to insects and diseases, (9) better protein quantity and quality, (10) crop canopies that can retain and fix a maximum of CO_2, and (11) suitability to mechanization.

Q11. What are the differences between traditional and ideotype breeding?

Ans. The differences between traditional and ideotype breeding are presented in the Table 19.1.

TABLE 19-1. Differences between Traditional and Ideotype Breeding

Sl.No.	*Particulars*	*Traditional Breeding*	*Ideotype Breeding*
1	Criteria decided in advance	Objective	Conceptual mode
2	Basis of selection	Yield and some other traits.	Individual plant character
3	Characters included	Morphological and economic.	Morphological, physiological and biochemical.
4	Speed of method	Rapid	Slow
5	Method	Simple	Difficult
6	Advance decision of the phenotype of new variety	Not specified	Specified.

Q12. What are the factors affecting ideotypes?

Ans. There are several factors which affect development of ideal plant type. Ideotype differs based on crop species, cultivation practices, socio-economic conditions of farmers and economic use of the plant parts. These are briefly discussed below:

1. **Crop Species:** Ideotype differs from crop to crop. The ideotype of monocots significantly differs from those of dicots. In monocots, tillering is more important whereas in dicots branching is one of the important features of ideotype.
2. **Cultivation:** The ideotype also differs with regard to crop cultivation. The features of irrigated crops differ from that of rainfed crop. The rainfed crop needs drought resistance, fewer and smaller leaves to reduce water loss through transpiration. In dicots, indeterminate types are required for rainfed conditions, because indeterminate type can produce another flush of flowers if the first flush is affected by drought conditions.
3. **Socio-economic Condition of Farmers:** Socio-economic condition of farmers also determines crop ideotype. For example, dwarf *Sorghum* is deal for mechanical harvesting in USA, but it is not suitable for the farmers of Africa where the stalks are used for fuel or hut construction.
4. **Economic Use:** The ideotype also differs according to the economic use of the crop. For example, dwarf types are useful in *Sorghum* and pearl millet when the crop is grown for grain purpose. But when these crops are grown for fodder purpose, tall stature is desirable one. Moreover, less leafy types are desirable for grain purpose and more leafy genotypes for fodder purpose. The larger leaves are also desirable in case of fodder crop.

Q13. What are the steps in ideotype breeding?

Ans. Ideotype breeding consists of four important steps, *viz.* (1) development of conceptual theoretical model, (2) selection of base material, (3) incorporation of desirable characters into single genotype, and (4) selection of ideal or model plant type. These steps are briefly discussed below:

1. **Development of Conceptual Model:** Ideotype consists of various morphological and physiological traits. The values of various morphological and physiological traits are specified to develop a conceptual theoretical model. For example, values for plant height, maturity duration, leaf size, leaf number, angle of leaf, photosynthetic rate *etc.*, are specified. Then efforts are made to achieve this model.
2. **Selection of Base Material:** Selection of base material is an important step after development of conceptual model of ideotype. Genotypes to be used in devising a model plant type should have broad genetic base and wider adaptability (Blixt and Vose, 1984) so that the new plant type can be successfully grown over a wide range of environmental conditions with stable yield. Genotypes for plant stature, maturity duration, leaf size and angle are selected from the global gene pool of the concerned crop species. Genotypes resistant or tolerant to drought, soil salinity, alkalinity, diseases and insects are selected from the gene

pool with the cooperation of physiologist, soil scientist, pathologist and entomologist.

3. **Incorporation of Desirable Traits:** The next important step is combining of various morphological and physiological traits from different selected genotypes into single genotype. Knowledge of the association between various characters is essential before starting hybridization programme, because it helps in combining of various characters. Linkage between desirable and undesirable traits hinders the progress of ideotype breeding. Various breeding procedures, *viz.* single cross, three way cross, multiple cross, backcross, composite crossing, intermating, mutation breeding, heterosis breeding, *etc.*, are used for the development of ideal plant types in majority of field crops. Backcross technique is commonly used for transfer of oligogenic traits from selected germplasm lines into the background of an adapted genotype.
4. **Selection of Ideal Plant Type:** Plants combining desirable morphological and physiological traits are selected in segregating populations and intermated to achieve the desired plant type. Morphological features are judged through visual observations and physiological parameters are recorded with the help of sophisticated instruments. Screening for resistance to drought, soil salinity, alkalinity, diseases and insects is done under controlled conditions. This task is completed with the help of scientist from the disciplines of physiology, soil science, pathology and entomology. Finally, genotypes combining traits specified in the conceptual model are selected, multiplied, tested over several locations, and released for commerical cultivation.

Q14. What are practical achievements of ideotype breeding?

Ans. Ideotype breeding has significantly contributed to enhanced yields in cereals (wheat and rice) and millets (*Sorghum* and pearl millet) through the use of dwarfing genes, resulting in green revolution. Semidwarf varieties of wheat and rice are highly responsive to water use and nitrogen application and have wide adaptation. These qualities have made them popular throughout the world.

Spontaneous mutations have played significant role in designing new plant types in wheat and rice. The Norin 10 in wheat and Dee-geo-Woo-gen in rice are the sources of dwarfing genes. These sources of dwarfing genes were obtained as a result of spontaneous mutations. Several high yielding semidwarf varieties have been evolved in wheat and rice through the use of respective dwarf mutant. The Norin 10 dwarfing gene in wheat, the Dee-geo-woogen dwarfing gene in rice and the genic cytoplasmic male sterile systems in *Sorghum* and pearl millet laid the foundation of green revolution in Asia (Swaminathan, 1972). Thus, ideotype breeding has been more successful for yield improvement in cereals and millets than in other crops.

In rice, the improved plant type includes (1) erect, short and thick leaves, (2) dwarf stature (short and thick straw), (3) light leaf sheath, (4) high tillering capacity, (5) responsiveness to high levels of nitrogen, and (6) high harvest index. Examples of semidwarf varieties of rice are IR 8, IR 20, TN 1 *etc.* The Chinese variety Dee-geo-Woogen is the source of dwarfing gene in all these varieties. In wheat, the improved plant type included (1) short and stiff straw (culms), (2) insensitivity to photoperiods, (3) high response to nitrogen application, (4) high harvest index, and (5) resistance to different rusts. The semidwarf Japanese variety Norin 10 was used as source of dwarfness in the development of semi dwarf Mexican varieties of wheat.

In *Sorghum* and pearl millet, short statured hybrids have been developed through the use of dwarfing genes. The dwarf hybrids have made machine harvesting possible in these crops. In *Sorghum*, combine harvesting has reduced the labour requirement by 1/8. In cotton, as a result of high selection pressure for earliness, short stature and compactness in the past, there has been a gradual reduction in the overall plant size. The earlier varieties were late maturing, tall growing and spreading types leading to bushy appearance. The target shifted to development of varieties with medium height, medium maturity and semi spreading habit. Now major emphasis is to evolve varieties with sort duration, short stature and compact plant type.

Q15. What are merits of ideotype breeding?

Ans. Merits of ideotype breeding are presented as follows:

(i) Ideotype breeding is an effective method of enhancing yield through manipulation of various morphological and physiological crop characters. Thus it exploits both morphological and physiological variation.

(ii) In this method values of various morphological and physiological traits are specified and each character or trait contributes towards enhanced yield.

(iii) Ideotype breeding involves experts from the disciplines of plant breeding, physiology, biochemistry, entomology and plant pathology. Each specialist contributes in the development of model plants for traits related to his field.

(iv) Ideotype breeding is an effective method of breaking yield barriers through the use of genetically controlled physiological variation for various characters contributing towards higher yield.

(v) Ideotype breeding provides solution to several problems at a time like disease, insect and lodging resistance, maturity duration, yield and quality by combining desirable genes for these traits from different sources into a single genotype.

(vi) It is an efficient method of developing cultivars for specific situation or environment.

Q16. What are demerits of ideotype breeding?

Ans. Demerits of ideotype breeding are presented as follows:

(i) Incorporation of several desirable morphological and physiological and disease resistance traits from different sources into a single genotype is a difficult task. Sometimes, combining of some characters is not possible due to tight linkage between desirable and undesirable characters. Presence of such linkage hinders the progress of ideotype breeding.

(ii) Ideotype breeding is a slow method of cultivar development, because combining together of various morphological and physiological features from different sources takes more time than traditional breeding where improvement is made in yield and one or two other characters.

(iii) Ideotype breeding is not a substitute for traditional or conventional breeding. It is a supplement to the former.

(iv) Ideotype is a moving object which changes with change in knowledge, new requirements, national policy, *etc.* Thus new ideotypes have to be evolved to meet the changing and increasing demands of economic products.

Q17. What are future prospects of ideotype breeding?

Ans. In India, the future research on crop ideotypes should be directed towards following aspects:

(i) India has achieved self-sufficiency in the production of food grains through modification of plant characters and development of high yielding varieties/hybrids. The further breakthrough in yield and quality has to be achieved through the exploitation of physiological variation. Ideotypes both for high and low input technology conditions have to be developed.

(ii) To further enhance the yield potential of food grain crops, ideotypes have to be evolved for straight varieties and hybrids. There is ample scope of developing hybrid ideotypes in crops like maize. *Sorghum*, pearl millet and rice. China has developed hybrid rice for commercial cultivation which covers more than 18 million hectares (Barwale, 1993).

(iii) Crop ideotypes have been developed especially in cereals and millets. There is ample scope for developing ideal plants or model plants in pulses, oilseeds, cotton and several other field crops. In these crops, again ideotypes have to be evolved both for irrigated as well as rain-fed cultivation. In cotton, ideotypes have to be developed for north, central and south zones which differ significantly from each other with regard to agro-climatic conditions.

(iv) In addition to traditional breeding approaches, biotechnological approaches, especially tissue culture and protoplast technology, have to be utilized in future for designing new plant types. Biotechnology may help in the development of insect resistant cultivars through the use of transgenic plants.

(v) Development of crop ideotype is a continuous process. Ideotype is a moving goal which changes with advancement in knowledge, new requirements, change in economic policy, *etc.*

(vi) Ideotypes should be developed to adverse conditions such as heat cold, salinity and drought conditions.

Chapter 20

Breeding for Climate Change

Q1. Describe main features of climate change.

Ans. Climate change is any long-term significant change in the "average weather" of a region or the earth as a whole. Average weather may include average temperature, precipitation and wind patterns. It involves changes in the variability or average state of the atmosphere over durations ranging from decades to millions of years. Main points related to climate change are listed below:

1. Climate change has continued throughout the entire history of Earth.
2. These changes can be caused by dynamic processes on Earth, external forces including variations in sunlight intensity, and more recently by human activities.
3. Climate change is a gradual process which takes very long time [decades to millions of years.
4. In recent usage, especially in the context of environmental policy, the term "climate change" usually refers to changes in modern climate.
5. Climate includes the average temperature, amount of precipitation, days of sunlight, *etc.*
6. It is believed beyond doubt that human activity is contributing to the current rapid changes in the world's climate.

Q2. What are the factors associated with climate change?

Ans. Climate changes reflect variations within the Earth's atmosphere, processes in other parts of the Earth such as oceans and ice caps, and the effects of human activity. The external factors that can shape climate are often called climate forcing and include such processes as variations in solar radiation, the Earth's orbit, and greenhouse gas concentrations.

Weather is the day-to-day state of the atmosphere. On the other hand, *climate* refers to the average state of weather and is fairly stable and predictable. Climate includes the average temperature, amount of precipitation, days of sunlight, and other variables that might be measured at any given site.

Climate change is the result of several factors such as (i) glaciation, (ii) ocean variability, (iii) CO_2 concentration, (iv) plate tectonics, (v) solar variations, (vi) orbital variations, (vii) volcanism, (viii) fossil fuel, (ix) cement manufacture, (x) land use (irrigation and deforestation), (xi) live stock *etc.* Detailed discussion of all these factors is beyond the scope of the present discussion. According to a 2006 United Nations report, Livestock's Long Shadow, livestock is responsible for 18 per cent of the world's greenhouse gas emissions as measured in CO_2 equivalents. All the above factors singly or in combination significantly contribute to climate change.

The sun is the source of over 99 per cent of the heat energy in the climate system. Less than 1 per cent of the energy is provided by the gravitational pull of the Moon (manifested as tidal power), in addition to geothermal energy provided by the hot inner core of the Earth. The energy output of the sun, which is converted to heat at the Earth's surface, is an integral part of shaping the Earth's climate.

Q3. What are the effects of climate change?

Ans. The climate change has two types of effects, *viz.* direct effects and indirect effects. The direct effects of climate change include effect on (i) temperature, (ii) rainfall, (iii) sunlight, (iv) CO2 concentration, *etc.* As a result of change in these aspects in a region, there are indirect effects or change in the (i) pest scenario, (ii)disease situation , (iii) water availability, (v) day length, (vi) biodiversity and (vii) cropping pattern. The effects of climate change in these variables are briefly discussed below:

(i) Increase in Temperature: Increase in temperature will lead to lead to drought/water deficit conditions. It will promote the incidence of certain types of insects, diseases and weeds that were not prevailing earlier in a particular region. Extreme temperature expands the desert area. Higher temperatures are expected to improve or retard seed germination, plant growth and/or plant development, depending on the relative sensitivity or tolerance of crop genotypes. Differences in climatic conditions, such as increases in temperature, will have a direct impact on local insect vector populations. Reproduction, physical activity and metabolic rates would all be increased. These processes would act together to increase the efficiency of vectors and consequently virus spread. In future, the use of insecticides will become restricted and methods to control vectors will require constant innovation.

The severe climate change leads to death or sometimes extinction of some plants and animal species. For example, in the UK during the drought year of 2006 significant numbers of trees died or showed dieback on light sandy soils. In Australia, since the early 90s, tens of thousands of flying foxes (Pteropus) have died as a direct result of extreme heat. Water deficit summers can lead to more periods of drought, potentially affecting many species of animal and plant.

(ii) **Change in Rainfall Pattern:** Climate change predictions are for increased variability of rainfall in many regions worldwide, resulting in greater fluctuations in soil water regime. The climate change may lead to increase or decrease of rainfall over the average value of rainfall in a particular region. The increase in the average rainfall may lead to flood and water stagnation conditions, whereas decrease in average rainfall may cause drought or water deficit conditions. The presence of many of the pathogens in the environment correlates with seasonal temperature and their spread is aided by rainfall.

(iii) **Change in Days of Sunlight:** The change in the average value of sunlight days in a particular region will have effect on the thermo and photoperiod requirements of the plant species grown in that region. As a result, new set of crop varieties will have to be bred for such situations. Only thermo and photo insensitive varieties can be successful under such situations.

(iv) **Increase in CO_2:** The increase in atmospheric CO_2 concentration will have a positive effect on productivity, though in a crop genotype-dependent manner.

(v) **Effect on Biodiversity:** Climate change and biodiversity interact in many important ways. In some cases, life cycles of many wild plants and animals are closely linked to the passing of the seasons. If two species are interdependent (*e.g.* a wild flower and its pollinating insect), climatic changes can disturb their synchronization. For example, if one has a cycle dependent on day length and the other on temperature or precipitation. In principle, at least, this could lead to extinctions or changes in the distribution and abundance of species.

Several organizations such as Wildlife Trust, World Wide Fund for Nature, Birdlife International and the Audubon Society are actively monitoring and researching the effects of climate change on biodiversity and advance policies in areas such as landscape scale conservation to promote adaptation to climate change are being framed.

Q4. What are the problems associated with climate change?

Ans. Gradual change in the climate over a long period will lead to several new problems such as appearance of new insects, new diseases, new weeds, high temperature, drought conditions, change in rainfall *etc.* As a result of climate change, the following problems emerge.

1. Water scarcity or excess depending upon the change in the rainfall pattern.
2. Drought if there is low rainfall and increase in temperature.
3. Change in response to thermo and photoperiods depending upon the change in sunlight and day length.
4. Appearance of new insects, diseases and weeds in the changed climate.
5. Change in the adaptation of cultivars in the areas of climate change.
6. Change in the cropping pattern in the region of climate change.

Such types of changes will require new set of crop cultivars for successful crop production. To keep pace with the climate change, breeders have to make continuous concerted efforts. The climate change needs to be converted from a difficulty into an opportunity.

TABLE 20-1. Climate Change and its Effects

Sl.No.	*Climate Change*	*Anticipated Effects*
1	Rise in Temperature	Drought, desert expansion, insects, diseases, crop adaptation, *etc.*
2	Rainfall Pattern	More Rainfall: Flood situation, water stagnation, cropping pattern. Low Rainfall: Water deficit, drought
3	Sunlight	Thermo and photo period requirement
4	Wind velocity	Lodging of crop cultivars

Q5. What are the sources which can be used in breeding for climate change?

Ans. In crop plants, there are five important sources which can be used for developing suitable cultivars for cultivation in the region where climate change is taking place. The sources include (i) germplasm collection, (ii) cultivated varieties, (iii) wild species and relatives of crop plants, (iv) induced mutations, and (v) trans-genes. These are briefly discussed below:

(i) **Germplasm Collections:** Germplasm Collections are the potential sources that can be effectively used in developing suitable crop cultivars for areas of climate change. The new cultivar has be developed, to solve the specific problem or a set of problems, using appropriate breeding techniques.

(ii) **Cultivated Varieties:** Crop cultivars can also be developed using obsolete or current crop varieties. Again the cultivar has to be developed to combat the problem of specific disease or insect or weed, or drought, or water stagnation that has emerged as a result of climate change. In the past, cultivated crop cultivars have been effectively used for developing varieties resistant to above mentioned changed situations.

(iii) **Wild Species and Relatives:** Wild Species and Relatives of cultivated plants are potential sources of developing crop cultivars suitable for climate change. However, use of these poses several problems such as cross in compatibility, hybrid in viability and hybrid sterility. Hence, such sources are rarely used.

(iv) Induced Mutations: Induced mutations are potential sources of developing crop cultivars suitable for climate change. In the past, induced mutations have been successfully used for developing varieties resistant to diseases and insects in several crop species.

(v) Transgenic Technology: Transgenic plants can be used to combat the problems arising due to climate change. Crop cultivars resistant to various insects and diseases can be developed through the use of transgenic technology.

Q6. What are the breeding approaches used for climate change?

Ans. Traditional breeding approaches will continue to develop cultivars suitable for climate change. These approaches include introduction, selection, hybridization and mutations. In developing suitable crop cultivars for climate change due importance should be given to: (i) development of MAR lines, (ii) genetic enhancement, (iii) transgenic technology, (iv) durable resistance, and (v) crop adaptation. These are briefly discussed below:

(i) Development of MAR Lines: Emphasis should be given to develop crop cultivars with multiple adversities resistance. This may include resistance to insects, diseases, drought and excessive moisture. This includes both breeding for multiple insect resistance [MIR] and breeding for multiple disease resistance [MDR].

(ii) Genetic Enhancement: This technique can be used in developing novel genetic diversity that can be used in breeding crop cultivars suitable for climate change.

(iii) Transgenic Breeding: Transgenic breeding can be successfully used for developing crop cultivars resistant to diseases, insects and drought conditions arising due to climate change.

(iv) Durable Resistance: Plant breeders should make use of long lasting resistance [durable resistance] for developing insect and disease resistant varieties. Multi-lines can be used for achieving such goal.

(v) Crop Adaptation: Maintaining food production under conditions of climate and environmental change will require the breeding of new crop varieties better adapted to these conditions. Heterogeneous crops also exhibit greater stability across environments. The best sources of pest and pathogen resistance can be further enhanced by deployment in such mixed crop stands.

For climate change, crop cultivars with wide adaptation should be selected based on multi-location under diversified environmental conditions for 3- years. In other words, crop adaptation should be given due importance in developing varieties suitable for climate change. Breeding strategies for different situations of climate change are presented in Table 20.2.

TABLE 20-2. Situation of Climate Change and Breeding Strategies

Sl.No.	*Situation of Climate Change*	*Breeding Strategies*
1	Low rainfall in paddy growing region.	(i)Development of upland paddy cultivars that can be grown with irrigation without standing water.(ii) Shifting of paddy with crops with low water requirement.
2	Very high rainfall in paddy growing areas.	Development of floating varieties of paddy.
3	Rise in temperature leading to drought situations.	Development of drought and heat resistant crop cultivars.
4	Sudden drop in temperature leading to snow fall, cold and frost conditions.	Development of crop cultivars resistant to freezing, cold and frost conditions.
5	Very high rainfall in pulses growing areas	Shifting of pulse crops with paddy and sugarcane.
6	Sudden appearance of new insects and diseases due to climate change.	Development of crop cultivars resistant to new insects and diseases.
7	Moisture deficit in wheat growing areas.	Development of drought resistant cultivars of wheat.
8	Low intensity of light and lesser sunlight days.	Development of thermo and photo insensitive crop cultivars.
9	Delayed on set of Monsoon	Development of crop varieties suitable for late sowing.
10	Increase in wind velocity	Development of crop cultivars resistant to lodging.

Q7. What are the screening techniques used for climate change?

Ans. The screening of breeding material can be carried out looking to the new problem arising as a result of climate change. The new problems may include heat, drought, excessive moisture, new insects, new diseases and new weeds. The standard procedures available cal be adopted for screening of material to tackle these problems. The material can be screened under both natural and controlled conditions to select genotypes suitable for a particular situation. Under artificial conditions the screening of breeding material should be carried out under controlled conditions representing climate change.

Q8. What are future breeding goals of climate change?

Ans. The climate change is occurring at the global level. As a result, India is facing a climate change of considerable magnitude. It is envisaged that as the twenty-first century progresses, there will be:

(i) Rise in temperature ranging from 2 to 6degree Celsius. The level of warming will be higher in the northern part of India than in the southern parts;

(ii) Rapid increase in night temperatures compared to day temperatures;

(iii) Increase in rainfall [15 to 40 per cent] over all the states, especially those in the western and central-west parts of India, excepting Punjab, Haryana, Delhi, Rajasthan and Tamil Nadu.

(iv) In view of the climate change in India, the future plant breeding efforts need to be directed towards the following thrust areas.

(1) The new crop varieties should be tolerant to drought and heat that are caused due to increase in temperature. Crop breeding programs to develop temperature and drought-tolerant high yielding cultivars of the identified crops should be initiated urgently, so that the desired kinds of crop cultivars are available when effects of climate change are noticed.

(2) The new varieties, especially grown during winter season should have characteristics of early flowering, photo- and thermo-insensitivity, early maturity and high productivity.

(3) The plant genetic resources, especially land races from those areas that have similarities with climate change can be used to start breeding programs for developing varieties suitable for climate change.

(4) The new crop cultivars for climate change can be developed using a combination of conventional breeding approaches, marker assisted selection, induced mutations and transgenic-breeding. Crop based coordinated programs need to be launched to develop early-maturing, high-yielding and temperature- and drought tolerant varieties as early as possible.

(5) The desirable genotypes for climate change can also be selected in the breeding populations of some ongoing research programs. There will be need for identification of areas where the climate change conditions already exist or resemble. In such areas, large segregating breeding populations can be screened for selection of desirable genotypes.

Q9. What are practical achievements related to climate change?

Ans. In the past, several cultivars in different crop species have been developed with resistance to various insects, diseases and drought. Conventional breeding approaches along with modern crop improvement techniques will be rewarding to solve various problems that arise in future as a result of climate change. Progress has been made in the genetic dissection of flowering time, inflorescence architecture, temperature and drought tolerance in certain model plant systems and by comparative genomics in crop plants. Recently, the Indian Agricultural Research Institute, New Delhi has released an early-maturing wheat variety suitable for late planting.

Appendices

Appendix 1. List of Agricultural Universities in India

Sl.No.	*Name of University*	*Location*	*State*
I. AGRICULTURAL UNIVERSITIES			
A	**STATE UNIVERSITIES**		
1	Acharya NG Ranga Agricultural University	Hyderabad	Telangana
2	Assam Agricultural University	Jorhat	Assam
3	Bihar Agricultural University	Sabour	Bihar
4	Birsa Agricultural University	Ranchi	Jharkhand
5	Indira Gandhi Krishi Vishwavidyalaya	Raipur	Chhatishgarh
6	Anand Agricultural University	Anand	Gujarat
7	Junagadh Agricultural University	Junagarh	Gujarat
8	Navsari Agricultural University	Navsari	Gujarat
9	Sardarkrushinagar-Dantiwada Agricultural University	Bansakantha	Gujarat
10	Chaudhary Charan Singh Haryana Agricultural University	Hisar	Haryana
11	CSK Himachal Pradesh Krishi Vishvavidyalaya,	Palampur	Himachal Pradesh
12	Sher-E-Kashmir Univ of Agricultural Sciences and Technology	Jammu	Jammu and Kashmir
13	Sher-E-Kashmir Univ of Agricultural Sciences and Technology of Kashmir	Srinagar	Jammu and Kashmir
14	University of Agricultural Sciences	Bangalore	Karnataka
15	University of Agricultural Sciences	Dharwad	Karnataka
16	University of Agricultural Sciences	Raichur	Karnataka
17	Kerala Agricultural University	Vallanikara	Kerela
18	Jawaharlal Nehru Krishi Viswavidyalaya	Jabalpur	Madhya Pradesh
19	Rajmata Vijayaraje Scindia Krishi Viswavidyalaya	Gwalior	Madhya Pradesh
20	Dr Balasaheb Sawant Konkan Krishi Vidyapeeth	Ratnagiri	Maharashtra
21	Dr Panjabrao Deshmukh Krishi Vidyapeeth	Akola	Maharashtra
22	Mahatma Phule Krishi Vidyapeeth	Rahuri	Maharashtra
23	Marathwada Agricultural University	Parbhani	Maharashtra

Sl.No.	*Name of University*	*Location*	*State*
24	Odisha Univ of Agriculture and Technology	Bhubaneswar	Odisha
25	Punjab Agricultural University	Ludhiana	Punjab
26	Maharana Pratap Univ of Agriculture and Technology	Udaipur	Rajasthan
27	Rajasthan Agricultural University	Bikaner	Rajasthan
28	Agriculture University	Kota	Rajasthan
29	Sri Karan Narendra Agriculture University	Jobner	Rajasthan
30	Agriculture University	Jodhpur	Rajasthan
31	Tamil Nadu Agricultural University	Coimbatore	Tamil Nadu
32	Chandra Shekar Azad University of Agriculture and Technology	Kanpur	Uttar Pradesh
33	Narendra Deva University of Agriculture and Technology	Faizabad	Uttar Pradesh
34	Sardar Ballabh Bhai Patel Univ of Agriculture and Technology	Meerut	Uttar Pradesh
35	Manyavar Kashiram University of Agriculture and Technology	Banda	Uttar Pradesh
36	Govind Ballabh Pant University of Agriculture and Technology	Pantnagar	Uttrakhand
37	Bidhan Chandra Krishi Viswavidyalaya	Kalyani	West Bengal
38	Uttar Banga Krishi Viswavidyalaya	Cooch Bihar	West Bengal
B	**CENTRAL UNIVERSITIES**		
1	Central Agricultural University	Imphal	Manipur
2	Rani Laxmibai Central Agricultural University	Jhansi	Uttar Pradesh
3	Rajendra Central Agricultural University	Pusa	Bihar
C	**DEEMED UNIVERSITIES**		
1	Indian Agricultural Research Institute	Pusa	New Delhi
2	National Dairy Research Institute	Karnal	Haryana
3	Indian Veterinary Research Institute	Bareilly	Uttar Pradesh
4	Central Institute for Fisheries Education	Mumbai	Maharashtra
5	Allahabad Agricultural Institute	Allahabad	Uttar Pradesh

II. ANIMAL SCIENCE UNIVERSITIES

S. N.	Name of University	Location	State
1	Sri Venkateswara Veterinary University	Tirupati	Andhra Pradesh

Sl.No.	Name of University	Location	State
2	Karnataka Veterinary, Animal and Fisheries Sciences University	Bidar	Karnataka
3	Maharashtra Animal Science and Fishery University	Nagpur	Maharashtra
4	Guru Angad Dev Veterinary and Animal Science University	Ludhiana	Punjab
5	Tamil Nadu Veterinary and Animal Science University	Chennai	Tamil Nadu
6	Tamil Nadu Fisheries University	Nagapattinam	Tamil Nadu
7	UP Pandit Deen Dayal Upadhaya Pashu Chikitsa Vigyan Vishwa Vidhyalaya evam Go Anusandhan Sansthan, Mathura	Mathura	Uttar Pradesh
8	West Bengal University of Animal and Fishery Sciences	Kolkata	West Bengal
9	Rajasthan University of Veterinary and Animal Sciences	Bikaner	Rajasthan
10	Kerala University of Fisheries and Ocean Studies	Kochi	Kerala
11	Kerala Veterinary and Animal Sciences University	Wayanand	Kerala
III. HORTICULTURAL UNIVERSITIES			
1	Andhra Pradesh Horticultural University	Tadepalli-gudem	Andhra Pradesh
2	University of Horticultural Sciences	Bagalkot	Karnataka
3	Dr Yashwant Singh Parmar Univ of Horticulture and Forestry	Solan	Himachal Pradesh
4	Dr YSR Horticultural University	Veketara-manna-gudem	Andhra Pradesh
5	Uttarakhand University of Horticulture and Forestry	Bharsar	Uttrakhand

Appendix 2. List of ICAR Research Institutes in India

Sl.No.	Name of Research Institute	Location	State
A	CROP SCIENCES		
1	ICAR-Central Institute for Cotton Research	Nagpur	Maharashtra
2	ICAR- Central Institute for Jute and Allied Fibres	Kolkata	West Bengal
3	ICAR-National Rice Research Institute	Cuttack	Odisha
4	ICAR -Central Tobacco Research Institute	Rajamundry	Andhra Pradesh
5	ICAR- Directorate of Groundnut Research	Junagarh	Gujarat
6	ICAR-Indian Institute of Oilseeds Research	Hydereabad	Telangana
7	ICAR- Directorate of Rapeseed-Mustard Research	Bharatpur	Rajasthan
8	ICAR-Indian Institute of Rice Research	Hyderabad	Talangana
9	ICAR-Indian Institute of Seed Science	Mau	Uttar Pradesh
10	ICAR-Indian Institute of Millets Research	Hyderabad	Talangana
11	ICAR-Indian Institute of Soybean Research	Indore	Madhya Pradesh
12	ICAR-Indian Agriculture Research Institute	Pusa	New Delhi
13	ICAR-Indian Agriculture Research Institute	Hazaribagh	Jharkhand
14	ICAR-Indian Grassland and Fodder Research Institute	Jhansi	Uttar Pradesh
15	ICAR-Indian Institute of Maize Research	Ludhiana	Punjab
16	ICAR-Indian Institute of Agricultural Biotechnology	Ranchi	Bihar
17	ICAR-Indian Institute of Pulses Research	Kanpur	Uttar Pradesh
18	ICAR-Indian Institute of Sugarcane Research	Lucknow	Uttar Pradesh
19	ICAR-Indian Institute of Wheat and Barley Research	Karnal	Haryana
20	ICAR-National Bureau of Agricultural Insect Resources	Bangaluru	Karnataka

Sl.No.	*Name of Research Institute*	*Location*	*State*
21	ICAR-National Bureau of Agriculturally Important Micro-organism	Mau	Uttar Pradesh
22	ICAR-National Bureau of Plant Genetic Resources	Pusa	New Delhi
23	ICAR-National Centre for Integrated Pest Management	Pusa	New Delhi
24	ICAR- National Institute of Biotic Stress Management	Raipur	Chhatishgarh
25	ICAR-National Centre on Plant Biotechnology	Pusa	New Delhi
26	ICAR-Sugarcane Breeding Institute	Coimbatore	Tamil Nadu
27	ICAR-Vivekanand Parvatiya Krishi Anusandhan Sansthan	Almora	Uttrakhand
B	**HORTICULTURAL SCIENCES**		
28	ICAR-Central Institute for Arid Horticulture	Bikaner	Rajasthan
29	ICAR-Central Institute for Subtropical Horticulture	Lucknow	Uttar Pradesh
30	ICAR-Central Institute for Temperate Horticulture	Srinagar	Jammu and Kashmir
31	ICAR- Central Island Agricultural Research Institute	Portblair	Andman and Nicobar
32	ICAR-Central Plantation Crops Research Institute	Kasargod	Kerala
33	ICAR-Central Potato Research Institute	Shimla	Himachal Pradesh
34	ICAR-Central Tuber Crops Research Institute	Trivencrum	Kerala
35	ICAR-Directorate of Cashew Research	Puttur	Karnataka
36	ICAR-Directorate of Floriculture Research	Pusa	New Delhi
37	ICAR-Directorate of Medicinal And Aromatic Plants Research	Anand	Gujarat
38	ICAR-Directorate of Mashroom Research	Solan	Himachal Pradesh
39	ICAR-Directorate of Onion and Garlic Research	Pune	Maharashtra
40	ICAR-Indian Institute of Horticulture Research	Bangalore	Karnataka

Sl.No.	Name of Research Institute	Location	State
41	ICAR-Indian Institute of Oil Palm Research	Pedavegi	Andhra Pradesh
42	ICAR-Indian Institute of Spices Research	Calicut	Kerala
43	ICAR-Indian Institute of Vegetable Research	Varanasi	Uttar Pradesh
44	ICAR-National Research Centre for Banana	Tiruchirapalli	Kerala
45	ICAR-Central Citrus Research Institute	Nagpur	Maharashtra
46	ICAR-National Research Centre for Grapes	Pune	Maharashtra
47	ICAR-National Research Centre for Litchi	Muzaffarpur	Bihar
48	ICAR-National Research Centre for Orchids	Gangtok	Sikkim
49	ICAR-National Research Centre on Pomegranate	Solapur	Maharashtra
C	**NATURAL RESOURCE MANAGEMENT**		
50	ICAR-National Research Centre for Seed Spices	Ajmer	Rajasthan
51	ICAR-Central Agro-forestry Research Institute	Jhansi	Uttar Pradesh
52	ICAR-Central Arid Zone Research Institute	Jodhpur	Rajasthan
53	ICAR-Central Research Institute for Dry land Agriculture	Hyderabad	Andhra Pradesh
54	ICAR-Central Soil and Water Conservation	Dehradun	Uttrakhand
55	ICAR-Central Soil Salinity Research Institute	Karnal	Haryana
56	ICAR-Indian Institute of Water Management	Bhubaneswar	Odisha
57	ICAR-Directorate of Weed Research	Jabalpur	Madhya Pradesh
58	ICAR Research Complex for NEH Region	Shillong	Meghalaya
59	ICAR Research Complex for Eastern Region	Patna	Bihar
60	ICAR-Indian Institute of Soil Science	Bhopal	Madhya Pradesh

Sl.No.	Name of Research Institute	Location	State
61	ICAR-Indian Institute of Farming System Research	Meerut	Uttar Pradesh
62	ICAR-Indian Institute of Soil Science	Bhopal	Madhya Pradesh
63	ICAR-National Bureau of Soil Survey and Land Use Planning	Nagpur	Maharashtra
64	ICAR-National Institute of Abiotic Research management	Baramati	Maharashtra
65	ICAR- National Organic Farming Research	Gangtok	Sikkim
66	ICAR_ National Research Centre for Integrated Farming	Morihari	Bihar
D	**AGRICULTURAL ENGINEERING**		
67	ICAR-Central Institute for Research on Cotton Technology	Mumbai	Maharashtra
68	ICAR-Central Institute of Agricultural Engineering	Bhopal	Madhya Pradesh
69	ICAR-Central Institute Post Harvest Engineering and Technology	Ludhiana	Punjab
70	ICAR-Indian Institute of Natural Resigns and Gums	Ranchi	Jharkhand
71	ICAR-National Institute of Research on Jute and Allied Fibres Technology	Kolkata	West Bengal
E	**AGRICULTURAL EDUCATION**		
72	ICAR-Central Institute for Women in Agriculture	Bhubaneswar	Odisha
73	ICAR-Indian Agricultural Statistics Research Institute	Pusa Campus	New Delhi
74	ICAR-National Academy of Agricultural Research Management	Hyderabad	Telangana
75	ICAR-National Institute of Agricultural Economics and Policy Research	Pusa Campus	New Delhi
F	**AGRICULTURAL EXTENSION**		
76	ICAR-Directorate of Knowledge Management in Agriculture	Pusa Campus	New Delhi
G	**ANIMAL RESEARCH INSTITUTES**		
77	ICAR-Central Avian Research Institute	Bareilly	Uttar Pradesh
78	ICAR-Central Institute for Research on Buffaloes	Hisar	Haryana

Sl.No.	Name of Research Institute	Location	State
79	ICAR-Central Institute for Research on Cattle	Meerut	Uttar Pradesh
80	ICAR-Central Institute for Research on Goats	Mathura	Uttar Pradesh
81	ICAR-Central Sheep and Wool Research Institute	Avikanagar	Rajasthan
82	ICAR-Directorate of Poultry Research	Hyderabad	Telangana
83	ICAR-Indian Veterinary Research Institute	Bareilly	Uttar Pradesh
84	ICAR-National Bureau of Animal Genetic Resources	Karnal	Haryana
85	ICAR-National Dairy Research Institute	Karnal	Haryana
86	ICAR-National Institute of Animal Nutrition and Physiology	Karnal	Haryana
87	ICAR-National Institute of High Security Animal Diseases	Bhopal	Madhya Pradesh
88	ICAR-National Institute of Veterinary Epidemiology and Disease Informatics	Bangalore	Karnataka
89	ICAR-National Research Centre for Equines	Hisar	Haryana
90	ICAR-National Research Centre for Camel	Bikaner	Rajasthan
91	ICAR-National Research Centre on Meat	Hyderabad	Telangana
92	ICAR-National Research Centre on Mithun	Jharnapani	Nagaland
93	ICAR-National Research Centre on Pig	Guwahati	Assam
94	ICAR-National Research Centre on yak	Dirang	Arunachal Pradesh
95	ICAR-Project Directorate on Foot and Mouth Disease	Nainital	Uttrakhand
H	**FISHERIES INSTITUTES**		
96	ICAR-Central Island Fisheries Research Institute	Barrackpur	West Bengal
97	ICAR-Central Institute of Brackish water Aquaculture	Chennai	Tamil Nadu
98	ICAR-Central Institute Fisheries Education	Mumbai	Maharashtra
99	ICAR-Central Institute Fisheries Technology	Cochin	Kerala

Sl.No.	Name of Research Institute	Location	State
100	ICAR-Central Institute of Fresh Water Aquaculture	Bhubaneswar	Odisha
101	ICAR-Central Marine Fisheries Research Institute	Kochi	Kerala
102	ICAR-Directorate of Cold Water Fisheries Research	Bhimtal	Uttarakhand
103	ICAR-National Bureau of Fish Genetic Resources	Lucknow	Uttar Pradesh

Appendix 3: List of International Agricultural Research Institutes/Centres

Sl. No.	*Name of Research Institute/Centre*	*Founded in*	*Location*	*Country*
1.	International Rice Research Institute [IRRI]	1960	Manila	Philippines
2.	International Wheat and Maize Improvement Centre [CIMMYT]	1963	Mexico City	Mexico
3.	International Centre for Tropical Agriculture [CIAT]	1967	Cali	Colombia
4.	International Institute for tropical Agriculture [IITA]	1967	Ibadan	Nigeria
5.	International Potato Centre [CIP]	1971	Lima	Peru
6.	International Crop Research Institute for Semiarid Tropics [ICRISAT]	1972	Hyderabad	India
7.	Internationa Centre for Agriculture Research in Drylans Areas [ICARDA]	1977	Aleppo	Syria
8.	West Africa Rice Development Association [WARDA]	1970	Bouake	Cote divoire
9.	International Plant Genetic Resources Institute [IPGRI]	1974	Rome	Italy
10.	Centre for International Forestry Research [CIFOR]	1992	Jakarta	Indonesia
11.	International Centre for Research in Agroforestry [ICRAF]	1991	Nairobi	Kenya
12.	International Centre for Living Aquati Resources Management [ICLARM]	1977	Makati City	Philippines
13.	International Food Policy Research Institute [IFPRI]	1975	Washington DC	USA
14.	International Irrigation Management Institute [IIMI]	1984	Colombo	Sri Lanka
15.	International Livestock Research Institute [ILRI]	1995	Nairobi	Kenya
16.	International service for National Agricultural research [ISNAR]	1979	Hague	Netherlands

Sl. No.	*Name of Research Institute/Centre*	*Founded in*	*Location*	*Country*
17.	Asian Vegetable Research and Development Centre [AVRDC]	1971	Taiwan City	Taiwan [China]
18.	International Centre for Genetic Engineering and Biotechnology [ICGEB]	1987	Trieste	Italy

Glossary

Chapter 1: Introduction and Objectives

Abiotic Stress: Adverse conditions imposed for growth and production of crop plants by environmental factors such as deficiency or excess of nutrition, moisture, temperature and light; presence of harmful gases or toxicants; and abnormal soil conditions, such as salinity, alkalinity and acidity.

Adaptability: Stable performance of a variety over a wide range of environmental conditions.

Biotic Stress: Adverse conditions imposed for growth and production of crop plants by biotic factors such as insects, diseases and parasitic weeds.

Breeding Techniques: Various breeding procedures which are used for genetic improvement of crop plants in relation to their economic use.

Easy Care Property: In cotton. fabrics that are easily washed and require little pressing.

Genetic Resistance: Ability of some genotypes to give higher yield of good quality than other varieties at the same initial level of disease or insect infestation under similar environmental conditions.

Germplasm: The whole library of alleles in a crop species or sum total of genes in a species.

Harvest Index: The ratio of economic yield to biological yield. Biological yield is the total dry matter production per plant.

Plant Breeding: A science, an art and a technology which deals with genetic improvement of crop plants in relation to their economic use for mankind, also called as crop improvement.

Quality: Suitability or fitness of an economic plant product in relation to its end use.

Seed Production Technology: A branch of plant breeding which deals with principles and methods of improved seed production.

Chapter 2: Centres of Origin and Distribution of Crop Species

Biodiversity: Total variability present within and among species of all living organisms and their habitats.

Centres of Diversity: A place, region or area where maximum variability of crop plants is observed; also called centres of origin.

Extinction: Permanent loss of a crop species due to various reasons.

Field Gene Banks: Those areas of land in which germplasm of recalcitrant crop species is maintained in the form of plants; also called plant gene banks.

Gene Banks: Various organizations where genetic diversity is maintained in living state; also called as germplasm banks.

Gene Sanctuaries: Protected areas of great genetic diversity under natural conditions *i.e. in situ* conservation.

Genetic Diversity: Total amount of genetic variation present in a population or species or variety of genes and genotypes found in a particular crop species.

Genetic Erosion: Loss of genetic diversity between and within population of a species over a period of time.

Law of Parallel Variation: This states that a particular variation observed in a crop species is also expected to be available in its another related species. Also called law of homologous series of variation.

Micro Centres: Small areas within the centre of diversity that exhibits tremendous genetic diversity of crop plants.

Plant genetic Resources: Genetic material of plants which is of value as a resource for present and future generations of people.

Primary Centre of Diversity: The original homes of crop plants which are generally uncultivated areas such as mountains, hills, river valleys, forests *etc.*

Secondary Centres of Diversity: The cultivated areas with vast genetic diversity of a species.

Seed gene banks: Conservation of germplasm in the form of seeds in cold storage.

Chapter 3: Plant Genetic Resources: Conservation and Utilization

Active Collections: Collections which are actively utilized in the breeding programmes and are used for medium term storage (8-10 years). Seeds are stored at 0°C.

Base Collections: Total accessions available in a crop. These are used for long storage (upto 100 years). Seeds are stored at -18 or -20°C; also called principal collections.

Characterization: Recording of highly heritable phenotypic characters.

Conservation: The protection of genetic diversity from genetic erosion either under natural conditions or by storing in gene banks.

Core Collections: A set of accession derived from base collections to represent the genetic spectrum in the whole collection.

Documentation: The process of compilation, analysis, classification, storage and distribution of information.

***Ex situ* Conservation:** Preservation of germplasm in the gene banks.

Exotic Collection: The germplasm which is collected or received from other countries.

Gene Pool: The whole library of different alleles of a species or sum total of genes in a species. Also called germplasm, genetic stock and genetic resources.

Genetic Erosion: Gradual reduction in genetic variability due to elimination of various geneotypes. It results due to use of modern cultivars, modernization of agriculture and various development activities.

***In situ* Conservation:** Conservation of germplasm under natural conditions.

Indigeneous Collection: The germplasm which is collected within the country.

Land Races. The primitive cultivars which were selected and cultivated by farmers for many generations. Land races have more genetic diversity, wider adaptability and high degree of resistance to biotic and abiotic stresses.

Modern Cultivars: The currently cultivated high yielding varieties.

Obsolete Cultivars: The improved cultivars of the recent past.

Orthodox Seeds: Seeds which can be dried to low moisture content and stored at low temperature without losing their viability.

Plant Exploration: Trips arranged for collection of germplasm from different areas.

Primary Gene Pool: The gene pool in which intermating is easy and leads to production of fertile hybrids. It includes genotypes of same species. It is designated as GP1.

Quarantine: The prophylactic measure that is used to prevent the entry of new diseases, insects and weeds from other countries.

Recalcitrant Seeds: Seeds which show very drastic loss in viability with decrease in moisture content below 12 to 13 per cent. Such species include coconut, mango, tea, coffee, rubber, jack fruit, oil palm *etc.* Such seeds cannot be conserved in seed banks.

Secondary Gene Pool: The genetic material that leads to partial fertility on crossing with primary gene pool. It includes genotypes of related species and is designated at GP2.

Tertiary Gene Pool: The genetic material which leads to production of sterile hybrids on crossing with primary gene pool. It is designated as GP3.

Working Collections: Collections which are frequently utilized by breeders in their crop improvement programs. These are stored for short term (3-5 years) at 5-10°C.

Chapter 4: Genetics of Qualitative and Quantitative Characters

Additive variance: Average effect of genes on all segregating loci. Such genes show lack of dominance, *i.e.*, intermediate expression.

Additive x Additive Epistasis: Interaction between two or more loci each exhibiting lack of dominance individually. It is denoted as *AA* and is fixable.

Additive x Dominance Epistasis: Interaction between two or more loci, one exhibiting lack of dominance and the other dominance. It is denoted as *AD* and is non-fixable.

Character: Any property of an individual showing heritable variation. It includes morphological, physiological, biochemical and behavioural properties.

Dominance Variance: Deviation from the mean value due to intra-allelic interaction.

Dominance x Dominance Epistasis: Interaction of two or more loci, each exhibiting dominance individually. It is represented as *DD* and is nonfixable.

Environmental Variation: The non-heritable variation which is entirely due to environmental effects and varies under different environmental conditions.

Epistatic Variance: Deviation from the mean value as a consequence of non-allelic interaction, *i.e.*, interaction between two or more genes. It is of three types *viz.*, additive x additive, additive x dominance and dominance x dominance.

Genetic Variance: Heritable portion of total or phenotypic variance. It is of three types, *viz.* additive, dominance and epistatic variances.

Genotypic Variation: The inherent or genetic variation which remains unaltered by environmental changes.

Oligogenic Trait: Characters which are controlled by one or few genes each having detectable individual effect and exhibit discontinuous variation.

Phenotypic Variation: The total variability which is observable.

Polygenic Traits: Characters which are governed by several genes each having small individual effect and exhibit discontinuous variation.

Chapter 5: Genetic Basis of Breeding Methods

Allogamy: Development of seed by cross pollination.

Autogamy: Development of seed by self pollination.

Heterogeneous Population: A population that is composed of genetically dissimilar plants such as land races, mass selected populations, composites, synthetics and multilines.

Heterozygous Population: A Population that segregates on selfing such as *F*1 hybrids, composites and synthetics. Heterozygous individuals have unlike alleles at the corresponding loci.

Homogeneous Population: A population of genetically similar plants such as pure line, *F*1 between two pure lines and progeny of a clone.

Homozygosity: The proportion of homozygous individuals in a segregating population. Homozygosity = $[(2m-1)/2m]n$, where m = number of generations of self pollination and n = number of gene pairs segregating.

Homozygous Population: A true breeding population such as pure lines and mass selected populations in self-pollinated species. Homozygous individuals have identical alleles at the corresponding loci.

Plant Breeding Methods: Various procedures (Selection, hybridization, mutation, *etc.*) that are used for genetic improvement of crop plants.

Chapter 6: Breeding Self Pollinated Crops [Introduction and Pure line Selection]

Acclimatization: Adaptation of an introduced variety to the new environment.

Artificial Selection: The selection made by human. It fevours those characteristics of plants that are related to yield and quality.

Characterization: Recording observations on highly heritable traits such as colour, shape, hairiness *etc.*

Cyclic Selection: Selection in one direction for one generation or season and in opposite direction in next generation or season.

Direct Introduction: Introductions which are immediately adapted to the changed environment.

Directional Selection: Selection in favour of an extreme phenotype such as earliness or lateness, dwarfness or tallness.

Disruptive Selection: Selection in favour of two extreme types, *viz.* earliness and lateness or tallness and dwarfness.

Domestication. The process of bringing wild and weedy species under human management.

Exotic Variety: A foreign variety which is directly recommended for commercial cultivation.

Indirect Introduction: Introductions which take some years for adaptation in changed environment.

Natural Selection: Selection that operates in nature without human interference. It favours those plant characters that are essential for survival (adaptation) of a species.

Plant Introduction: Transposition of crop plants from the place of their cultivation to such areas where they were never grown earlier.

Primary introduction: Introduction that can be used for commercial cultivation without any change in the original genotype.

Pure line Selection: Development of new variety through identification and isolation of a single best plant progeny.

Pure Line: Progeny of a self-pollinated homozygous plant obtained by selfing.

Secondary Introduction: Introduction that can be used as a variety after selection from original genotype or used for transfer of some desirable gene to the cultivated variety.

Selection: The process that favours survival and further propagation of some plants having more desirable characters than others. It is of two types, *viz.*, natural and artificial.

Stabilizing Selection: Selection in favour of intermediate types; also called centripetal selection.

Chapter 7: Breeding Self Pollinated Crops [Pedigree, Bulk, Backcross and SSD Methods]

Backcross: Crossing of F_1 with either of its parents. When F_1 is crossed with homozygous recessive parent, it is called test cross.

Bulk Breeding: A selection procedure which is used in segregating population of self-pollinated species in which material is grown in bulk plot from F_2 to F_5 with or without selection, next generation is grown from bulk seed and individual plant selection is practiced in F_6 or later generations.

Donor Parent: The parent which donates desirable genes; also called non-recurrent parent, because it is used once in crossing programme.

Hybridization: Crossing between genetically dissimilar plants. It may involve two genotypes of the same species (inter-varietal hybridization) or two species of the same genus (interspecific hybridization) or two genera of the same family intergeneric hybridization).

Isogenic Lines: Genotypes having single locus difference only.

Mass Pedigree Method: A modified form of pedigree method in which segregating material is handled by bulk (mass) method when conditions are unfavourable for selection and by pedigree method when conditions are favourable for selection.

Multi-lines: The deliberate seed mixtures of isogenic lines, closely related lines or unrelated lines. A variety developed for commercial cultivation from any of these mixtures is known as multiline variety and such procedure is called multiline breeding.

Pedigree Breeding: A selection procedure which is used in segregating population of self-pollinated species and keeps proper record of plants and progeny selected in each generation.

Pedigree: Record of the ancestry of an individual selected plant for its various generations.

Recipient Parent: The parent which receives a desirable character, also called recurrent parent, because it is repeatedly used in backcrossing.

Single Seed Descent: A breeding procedure which is used with segregating population of self-pollinated species in which plants are advanced by selecting single seed per plant from F_2 generation onwards.

Chapter 8: Breeding Cross-Pollinated Crops [Mass Selection and Progeny Selection]

Line Breeding: A system of breeding in which a number of genotypes with superior performance for several characters are composed to form a variety.

Mass Selection: A method of crop improvement in which individual desirable plants are selected on the basis of phenotype from a mixed population, their seeds are bulked and used to grow the next generation **(positive mass selection).** Sometimes, only undesirable off type plants are removed from the field and rest are allowed to grow further **(negative mass selection).**

Progeny Selection: A selection procedure in which superior plants are selected from a heterogeneous population on the basis of the performance of their progeny.

Chapter 9: Breeding Vegetatively Propagated Crops [Clonal Selection]

Clonal Selection: A procedure of selecting superior clones from the mixed population of asexually propagated crops such as sugarcane, potato, *etc.*

Clone: Progeny of a single plant obtained by asexual reproduction.

Chapter 10: Transgenic Breeding

Biotechnology: Application of various biological organisms or processes for mass production of useful substances or products for industry, medicine and agriculture.

Gene Revolution: Quantum jump in the productivity of various field crops through the use of transgenic cultivars.

Protoplasts: Naked cells or cells without cell wall.

Transgenes: Foreign genes or modified genes of the same species which are used for development of transgenic individuals.

Transgenic Breeding: Genetic improvement of crop plants, domestic animals and useful micro-organisms, through application of biotechnology, in relation to their economic use for mankind.

Transgenic: Individuals containing altered or transgenes from an unrelated organism. It involves taking genes from one species and inserting them into another species to get that trait expressed in the offspring.

Chapter 11: SMART Breeding [Marker Assisted Selection]

Biochemical Marker: Markers that are related to variation in proteins and amino acid banding pattern.

Cytological Marker: Markers that are related to variation in chromosome number, shape and size.

DNA Marker: A gene or other fragment of DNA whose location in the genome is known. A genetic marker is a known DNA sequence.

Epistasis: The interaction between genes. Epistasis takes place when the action of one gene is modified by one or several other genes.

Genotyping: The process of determining the genotype of an individual with a biological assay

Locus (plural loci): A fixed position on a chromosome, such as the position of a gene or a biomarker (genetic marker).

Marker assisted selection: Selection for specific alleles (which affect a trait of interest) using genetic markers.

Marker: Any genetic element (locus, allele, DNA sequence or chromosome feature) which can be readily detected by phenotype, cytological or molecular techniques, and used to follow a chromosome or chromosomal segment during genetic analysis.

Molecular Breeding: Improvement of crop plants for various economic characters through indirect selection for linked molecular markers.

Morphological Marker: In plant breeding, markers that are related to variation in shape, size, colour and surface of various plant parts.

Phenotype: Any observable characteristic of an organism, such as its morphology, development, biochemical or physiological properties, or behavior.

Pleiotropy: Influence of a single gene on multiple phenotypic traits.

Polymorphism: Presence of multiple alleles of a gene within a population, usually expressing different phenotypes

Chapter 12 and 13: Varietal and Hybrid Seed Production

Breeder Seed: The progeny of nucleus seed or breeder seed produced under the strict supervision of original or sponsoring plant breeder.

Certified Seed: The progeny of either foundation or registered or certified seed. The certification is done by the State Seed Certification Agency.

Foundation Seed: The progeny of breeder seed produced by National Seeds Corporation.

Genetic Purity: The absence of seeds of other variety of the same crop as well as of other crops.

Germination: Emergence of normal seedlings from the seeds under ideal conditions of light, temperature, moisture, oxygen and nutrients.

Inert matter: Non- living materials such as sand, pebbles, soil particles, straw, *etc.*

Isolation: Separation of the field of a variety from that of another variety of the same crop to prescribed standard distance to avoid contamination.

Nucleus Seed: The initial seed of an improved variety, limited in quantity and produced by originating plant breeder.

Physical Purity: Freedom of seed from inert matter and defective seeds.

Registered Seed: The progeny of either foundation or registered seed. In India, this category is generally omitted and certified seed is produced directly from foundation seed.

Rogueing: The process of removal of off type (phenotypically different) plants from the field of an improved variety to avoid contamination.

Seed Certification: A legal system which ensures production of high quality seed in terms of genetic purity and germination.

Seed testing: The process of evaluation of seeds in terms of purity and germination.

Seed: In broad sense, any plant part which is used for commercial multiplication of a crop. In strict sense, the product of fertilized ovule that consists of embryo, seed coat and cotyledon(s).

Varietal Deterioration: Permanent reduction either in the genetic or agronomic value of a released variety.

Variety: A genotype released for commercial cultivation either by State Variety Release Committee or Central Variety Release Committee.

Chapter 14-18: Crop Breeding

Bagasse: In sugarcane, the fibrous residue which is left over crushing of cane.

Cereal Crops: The starchy grain of various grasses such as wheat, oats, or corn, which are used as food refer to cereal crops.

Cotton: Those species of the genus *Gossypium* which produce sinnable seed coat fibres.

Crop Plants: Domesticated and commercially cultivated plants or agricultural produce, such as grain, vegetables, or fruit, considered as a group are called crop plants: Wheat is a common crop.

Fibre Crops: Field crops which are grown for their fibers, which are traditionally used to make paper, cloth, or rope are called fibre crops. Such crops include cotton, Jute, sunnhamp, Kenef, *etc.*

Field Crop: A crop (other than fruits or vegetables) that is grown for agricultural purposes is called field crop. Cotton, pulses, sugarcane wheat, mustard, maize are examples of field crops"

Food Crops: Crops that can be used for food for humans. Examples: wheat, corn, fruits, vegetables are known as food crops.

Ginning: In cotton, the process of separating lint from the seed.

Husk: In maize, leafy structures covering the ear.

Monopodia: In cotton, vegetative branches.

Nobilization: In sugarcane, the process of crossing with *Saccharum spontaneum* and back crossing with noble cane *i.e. S. officinarum.*

Peg: In peanut, a long stalk bearing pod at the apex; also known as gynopore or carpopodium.

Pulse Crops: Leguminous plants producing edible seeds, *e.g.* chickpeas, lentils, beans, black gram, green gram, pigeon pea *etc.* are known as pulse crops.

Pulses: Those species of the family Leguminosae whose seeds are used for human consumption.

Shell: In groundnut, the pericarp of the pod.

Silk: In maize, long stigmas which emerge from upper end of cob.

Square: In cotton, unopened floral bud.

Sympodia: In cotton, fruiting branches.

Tassel: In maize, the male inflorescence.

Test Weight: Weight of 100 seeds in gram.

Trash: The dried sugarcane leaves. In cotton, dried leaf bits or any other foreign matter in the lint.

Tuber Crops: A swollen, fleshy, usually underground stem of a plant, such as the potato, bearing buds from which new plant shoots arise is known as tuber crop.

Chapter 19: Ideotype Breeding

Biological yield: Total dry matter production per plant.

Harvest Index: The ratio of economic yield to the biological yield or the ratio of economic produce to the total biomass.

Ideotype: A plant model which is expected to yield greater quantity of grains, fibre, oil or other useful product when developed as a cultivar.

Ideotype breeding: A method of crop improvement which is used to enhance genetic yield potential through genetic manipulation of individual plant character.

Chapter 20: Breeding for Climate Change

Breeding for climate change: Plant breeding strategies used develop crop varieties for change in climate of a region.

Climate change: Any long-term significant change in the "average weather" of a region or the earth as a whole.

Weather: The day to day atmospheric conditions such as temperature, precipitation and wind patterns.

References

1. **ALLARD, R. W.** 1960. Principles of Plant Breeding. John Wiley and Sons Inc. New York.
2. **ALLARD, R. W.** 1999. Principles of Plant Breeding 2nd edition. John Wiley and Sons Inc. New York.
3. **BOROJEVIC, S.** 1990 Principles and Methods of Plant Breeding. Elsevier, Amsterdam.
4. **BRIGGS, F. N. AND KNOWLES, P. F.** 1967. Introduction to Plant Breeding. Reinhold, New York.
5. **CHOPRA, V. L. (ED.)** 2000. Plant Breeding: Theory and Practice 2nd edition, Oxford and IBH Publishing Company, Pvt. Ltd. New Delhi.
6. **FEHR, W. R.** 1987. Principles of Cultivar Development Volume 1. Theory and Techniques. Macmillan Publishing Company, New York.
7. **FREY, K. J.** (ed.) 1966. Plant Breeding. Iowa State University, Press, Ames, Iowa, USA.
8. **FREY, K. J.** (ed.) 1981. Plant Breeding II. Iowa State University, Press, Ames, Iowa, USA.
9. **HARTEN, A. V.** 1998. Mutation Breeding: Theory and Practical Applications. Cambridge Uni. UK.
10. **KUCKKUCK, H., KOBADE, G. AND WENGEL G.** 1993. Fundamentals of Plant Breeding. Narosa Publishing House, New Delhi.
11. **MAYO, O. 1984**. The Theory of Plant Breeding. Clarendon, Oxford.
12. **POEHLMAN, J. M.** 1987. Breeding Field Crops 3rd edn.AVI Publishing Co. Inc. West Port, Connecticut, USA.
13. **POEHLMAN, J. M. AND BORTHAKUR, D. N.** 1969. Breeding Asian Field Crops. Oxford and IBH Publishing Company, New Delhi.

14. **PHUNDAN SINGH** 2018. Essentials of Plant Breeding 7th edition, Kalyani Publishers, New Delhi.
15. **PHUNDAN SINGH** 2016. Genetics 3rd edition, Kalyani Publishers, New Delhi.
16. **PHUNDAN SINGH** 2012. Cotton Breeding 3rd edition, Kalyani Publishers, New Delhi.
17. **PHUNDAN SINGH AND SANJEEV SINGH** 2010. Breeding Hybrid Cotton 3rd edn. Kalyani Publishers, New Delhi.
18. **SHARMA, J. R.** 1984. Principles and Practice of Plant Breeding. Tata McGraw-Hill Publishing Company Ltd. New Delhi.
19. **SIMMONDS, N. W.** 1979. Principles of Crop Improvement. Longman, London.
20. **SNEEP, J.AND HENDRIKSON, A. J. T.** (eds.) 1979. Plant Breeding Perspectives. Pudoc, Wageningen, Netherlands.
21. **VOSE, P. B. AND BLIXT, S. G.** (eds.) 1984. Crop Breeding A Contemporary Basis.Pergamon Press, London.
22. **WILLIAMS, W.** 1964. Genetic principles and Plant Breeding, Oxford, Blackwell.

Books by Dr. Phundan Singh

A. Plant Breeding

1. Essentials of Plant Breeding
2. Fundamentals of Plant Breeding
3. Plant Breeding: Molecular and New Approaches
4. Molecular Plant Breeding
5. Commercial Plant Breeding
6. Commercial Plant Breeding: At A Glance
7. Commercial Plant Breeding: Objective
8. Objective Science of Plant Breeding
9. Genetic Principles and Plant Breeding
10. Genetics and Plant Breeding: Comparative Analysis
11. Principles of Plant Breeding[For PG Students]
12. Plant Breeding: At a Glance
13. Plant Breeding [For Under Graduate Students]
14. Principles of Plant Breeding [For UG Students]
15. Practical in Crop Breeding
16. Practical and Numerical in Plant Breeding
17. Numerical Problems in Plant Breeding and Genetics
18. Glossary of Plant Breeding and Genetics
19. Breeding Crop Plants for Stress Resistance
20. Introduction to Resistance Plant Breeding

21. Resistance Plant Breeding: Objective
22. Resistance Plant Breeding: At A Glance
23. Plant Breeding: Related Legislations
24. Concepts in Plant Breeding
25. Principles of Seed Technology
26. Seed Technology: At a Glance
27. Objective Seed Technology

B. Cotton Breeding and Genetics

28. Cotton Breeding
29. Heterosis Breeding in Cotton
30. Breeding Hybrid Cotton
31. Breeding Transgenic Bt. Cotton
32. Basics of Bollgard Cotton
33. Basics of Bollgard Cotton: At A Glance
34. Basics of Bollgard Cotton: Objective
35. Cotton Improvement in India
36. Sankar Kapas [In Hindi]
37. Glimpses of Cotton Breeding
38. Cotton Genetics
39. Cotton

C. Genetics

40. Elements of Genetics
41. Genetics
42. Principles of Genetics [For UG Students]
43. Fundamentals of Genetics
44. Plant Genetics
45. Objective Genetics
46. Genetics: At a Glance
47. Objective Genetics and Plant Breeding
48. Objective Genetics and Plant Breeding [Hindi Edition]
49. Glossary cum Dictionary of Genetics and Plant Breeding
50. Molecular Genetics [Objective]
51. Molecular genetics [At a Glance]

52. Molecular Genetics [Subjective]
53. Genetics and Man
54. Genetics Today

D. Quantitative Genetics

55. Biometrical Techniques in Plant Breeding
56. Application of Biometrical Techniques in Plant Breeding [Hindi Edition]
57. Objective Quantitative Genetics
58. Quantitative Genetics: At a Glance
59. Quantitative Genetics
60. Introduction to Agricultural Statistics
61. Agricultural Statistics: Objective
62. Agricultural Statistics: At A Glance

E. Plant Biotechnology

63. Introduction to Biotechnology
64. Plant Biotechnology
65. Principles of Plant Biotechnology
66. Plant Biotechnology: At a Glance
67. Objective Plant Biotechnology
68. Fundamentals of Molecular Biology and Plant Biotechnology
69. Objective Molecular Biology and Plant Biotechnology
70. Molecular Biology and Plant Biotechnology: At A Glance

F. Intellectual Property Rights

71. IPR and Plant Breeders' Rights [Subjective]
72. IPR and Plant Breeders' Rights [Objective]
73. IPR and Plant Breeders' Rights [At a Glance]
74. Introduction to Intellectual Property Rights
75. Intellectual Property Rights: At A Glance
76. Intellectual Property Rights: Objective

G. Miscellaneous Books

77. Elements of Baby Corn
78. Baby Corn Farming for Higher Income
79. A Guide for Competitive Examinations of Agriculture
80. Objective General Agriculture for Competitive Examinations

81. Autobiography of Phundan Singh
82. Concepts in Plant Breeding [Objective Analysis]
83. Concepts in Plant Breeding [Question and Answer]
84. Principles of Plant Breeding [Objective Analysis]
85. Principles of Plant Breeding [Question and Answer]
86. Principles of Crop Improvement [Objective Analysis]
87. Principles of Crop Improvement [Question and Answer]